Injection Molding of Thermoplastics Materials – 2

Tony Whelan and John Goff

Extreme care has been taken in preparation of this work. However, neither the publisher nor the authors shall be held responsible or liable for any damages resulting in connection with, or arising from, the use of any of the information in this book.

Copyright © 1990 by Tony Whelan and John Goff

Library of Congress Catalog Card Number
ISBN 0-442-30550-8

All rights reserved. No part of this work covered by the copyright hereon may be reproduced or used in any form or by any means – graphic, electronic, or mechanical, including photocopying, recording, taping, or information storage and retrieval systems – without written permission of the publisher.

First published in the U.S.A. in 1990 by
Van Nostrand Reinhold
115 Fifth Avenue
New York, New York 10003

Van Nostrand Reinhold International Company Limited
11 New Fetter Lane
London EC4P 4EE, England

Van Nostrand Reinhold
480 La Trobe Street
Melbourne, Victoria 3000, Australia

Nelson Canada
1120 Birchmount Road
Scarborough, Ontario M1K 5G4, Canada

16 15 14 13 12 11 10 9 8 7 6 5 4 3 2 1

Library of Congress Cataloging-in-Publication Data
Whelan, Tony.
    Injection molding of thermoplastics materials, II/by Tony Whelan and John Goff.
      p.   cm.
    ISBN 0-442-30550-8
    1. Injection molding of plastics – Handbooks, manuals, etc. 2. Thermoplastics – Handbooks, manuals, etc. I. Goff, J.P. II. Title.

TP1150.W485 1990	89-78423
668.4'12–dc12	CIP
	Rev.

Preface

Over the years 1984 to 1989, we published a series of articles on the molding of thermoplastics, and of thermosetting materials, in the monthly magazine British Plastics and Rubber (B P & R). These articles were very well received and we also received a large number of requests for reprints. The articles were also translated into languages other than English.

In order to cater for what is obviously a need in both the thermoplastics, and the thermosetting, molding industries, we therefore brought the information together and produced it in book form. To make the material easier to handle we produced it in the form of several books and this is one of them. We can only hope that the information so presented, serves you well and that you find the information useful. We in turn would like to thank the editor of the magazine B P & R for helping us in this matter.

Thanks are also due to our many friends and colleagues throughout the molding industry for their useful help and advice: in particular, the company Moldflow (Europe) Limited deserve a special mention as they allowed us to extract information from their extensive data base.

The information presented in this publication is intended to give the reader an over view of the processing and properties of the materials listed in the index of this book. More detailed information, on specific grades or applications, should be obtained from the manufacturers of the material or, from the suppliers. We therefore suggest that, the user of such materials, seek and adhere to, the manufacturers or suppliers current instructions for the handling of each grade or type of material they use. Any determination of the suitability of the material for any use contemplated by the user, and the manner of processing and use, is the sole responsibility of the user who must satisfy himself/herself that the material as processed meets the needs of a particular product or use.

The information presented in this publication was based on our knowledge and experience and represents the opinions of us, the authors, and not necessarily the opinions of any other person or organization. To the best of our knowledge the information is accurate; however we do not assume any liability whatsoever for the accuracy or completeness of such information. Such information is given in good faith but in view of the many factors that affect processing and properties, the data so presented does not relieve processors from the responsibility of carrying out their own tests and experiments: neither does the data presented imply any legally binding assurance of certain properties, ease of processing or, of suitability for a given purpose. It is also not our responsibility to ensure that proprietary rights are not infringed and to ensure that relevant legislation is observed.

To the extent that any hazards may have been mentioned in this publication, we neither suggest nor guarantee that such hazards are the only ones which exist. We strongly recommend that anyone intending to rely on any suggestion or recommendation (or to use any equipment, processing technique or material mentioned in this publication) should satisfy himself/herself that all applicable safety and health standards can be met. Remember, there is a need to reduce human exposure to many organic materials to the lowest practical limits, in view of possible long-term adverse effects.

John Goff and Tony Whelan.

COMMODITY OR BULK MATERIALS

This group of materials covers the major materials such as polypropylene, polyvinyl chloride, polystyrene and polyethylene (PP, PVC, PS and PE). Each of these divisions covers a number of sub-divisions, for example PP covers homopolymers and copolymers, PVC covers rigid and plasticized, PS covers general purpose PS and toughened PS (TPS, or IPS, or HIPS) and PE covers LDPE, LLDPE and HDPE. The importance of bulk thermoplastics cannot be over-emphasized; approximately 70% of all plastics used fall into this category.

These materials are used so extensively because the raw materials on which they are based are relatively cheap and, conversion into plastics which have useful properties is fairly easy. It is also now being realized that the properties of this class of materials can be up-graded so that they compete with other more exotic, and therefore more expensive, plastics. A case in point is UPVC; because of the inherent flame resistance if this material, it has been developed so that it can now compete with engineering thermoplastics in some applications.

ENGINEERING THERMOPLASTICS

Engineering plastics are a group of polymers which offer a combination of some of the following, high strength, stiffness, toughness and, resistance to wear, chemical attack and heat.

Any discussion of engineering thermoplastics must concentrate on materials such as polyamides, acetals, polycarbonates, thermoplastic polyesters and modified polyphenylene oxide (also known as modified polyphenylene ether or PPE). Greater than 90% of all the engineering plastics used in the world, fall into one of these categories (approximately 8% of all plastics usage is accounted for by these five categories). Other materials such as the sulfones and the modified polyimides only account for a very small percentage (approximately 1 to 2%) of the total usage of engineering plastics. This relatively low consumption occurs, because most of the engineering plastics use, is in metal replacement where exotic property (and therefore expensive property) characteristics, are not required. The 'tonnage', engineering thermoplastics are used in a wide range of applications even though they have comparatively low heat resistance (below 150°C or 302°F) and relatively low creep resistance especially at high, or elevated, temperatures.

THERMOPLASTIC ELASTOMERS

Those materials which are referred to as 'thermoplastic elastomers' (TPE) or, as 'thermoplastic rubbers' (TPR) are a comparatively new group of materials but already they have become significant in terms of both tonnage and in terms of value. This is because they are creating new markets as well as replacing both plastics and traditional rubbers in some of their applications. They exhibit some of the properties associated with traditional, or vulcanized, rubbers, but they also exhibit the speed, and ease, of processing of thermoplastics; the ability to be re-processed is also important. Of major importance is the wide color range possible from this type of material. TPEs may be based on specifically manufactured polymers (e.g. styrene butadiene styrene or polyether ester elastomer - SBS and PEEL) or, they may be based on blends of rubbers and plastics. For example, a widely used material is that obtained by mixing PP with an ethylene propylene rubber. A two phase material results in which there is a plastics phase and a rubber phase. In the simplest cases, the rubber is not crosslinked or vulcanized but, is simply dispersed throughout the plastics material. There are other materials which also contain a dispersed rubber phase but, in this case, the rubber is partially vulcanized: these may be known as thermoplastics vulcanizates (TPV).

It is believed that the usage of TPEs will increase to such an extent that they will displace traditional rubbers from many applications. Such traditional elastomers/rubbers will continue to be used for applications requiring high temperature resistance and dynamic performance but TPEs will serve better for all other applications.

The reasons cited for the anticipated decline of traditional rubbers, in favor of TPEs, include the availability of a variety of TPEs of consistent quality, the elimination of customer formulation and compounding, easy and rapid processing, elimination of non-reusable scrap and, the often superior resistance of TPEs to chemical attack: it is often possible to replace a range of rubber mixes by one TPE because of the superior chemical resistance of the TPE.

BLENDS

There are a number of reasons for blending materials and these include the generation of unusual materials, extending the performance of existing plastics, extending this performance quickly, extending the performance cheaply and, re-using plastics materials. Both time and cost are against the development of new plastics. It may, for example, take fifteen years to develop a new plastic and cost many millions of dollars: a new blend can be produced for a fraction of this cost. The other factor is that new polymers, as envisaged by the manufacturers, have rarely satisfied the market for very long and, once original markets have developed, demands usually arise for performance improvements.

The blends which result are often produced on twin screw

extruders using what is called reaction compounding. As many twin screw machines are modular in design, their construction can be tailored to suit a particular resin blend thus facilitating production and giving the all-important, blend consistency. Because of the relatively poor performance of many engineering plastics in some applications, and/or because of their high price, recent attention has focussed on engineering polymer blends (EPB).

Engineering polymer blends (EPB) can be roughly divided into blends of an engineering plastic or resin (ER) with a commodity plastic, blends of an engineering plastic with another engineering plastic, blends of an engineering plastic with an elastomer and, blends which contain three or more polymers. We can therefore get combinations such as PPO/PS; PPO/PA; PC/ABS; PET/PBT: PBT/PC/SAN etc.; each of the blends may in turn be filled. When blends are made the objective is to simultaneously optimize blend formulations, in respect of several properties important for a particular application, sacrificing those which are not important.

THERMOSETTING PLASTICS

A thermosetting plastic, or a thermoset, may be defined as a material which once shaped by heat and pressure is incapable of being re-processed by further applications of heat and pressure. This is because during the shaping process, cross-links, or crosslinks, are introduced between the molecules which make up the basic resin or plastic. It is these cross-links which make the cured material insoluble and infusible as they stop the molecules slipping past each other and so prevent re-softening. This does not mean to say that these plastics are chemically inert: as will be seen, they are swollen by some chemicals and attacked by others.

Materials such as polyethylene may be cross-linked, after they have been shaped as a thermoplastic, but such materials will not be considered as thermosetting materials in these guides. Those materials which will be treated as thermosets are those which are traditionally known as thermosets - such as phenolics and aminoplastics: these materials are processed by the application of fairly high pressures in processes such as compression molding, transfer molding and injection molding. Also included in this category of materials are epoxides and, molding compositions based on an unsaturated polyester, for instance DMC.

ELASTOMERS

An elastomer (a rubber) may be defined as a material which may be stretched by 100% and on release of the stretching force it retracts, or springs back, rapidly. This elasticity was always associated with vulcanized (cross-linked) rubber and, is due to the structure of that type of material. Now, thermoplastics materials can be made rubber-like.

The long chain molecules on which rubbers are based, prefer to exist in a random, or coiled, configuration. However, the same, long chain molecules are not rigid structures but can untwist (or unwind) around the carbon to carbon bonds, on the application of a stretching force - provided the material is above its glass transition temperature. The chain is not stretched it is simply made to adopt a different configuration, or layout, but it always remembers that it would prefer to exist in a random, or coiled, state. So, when the stretching force is removed, back it goes to a random, coiled state. (A rubber molecule may be likened to a steel spring which, because of the way that it is constructed, may be easily stretched).

The crosslinks serve to stop the chains slipping past each other when they are stretched. If they were not there, then the chains would find it easier to untangle, orientate in the direction of force application, and then move past each other: this would cause the material to separate into pieces.

MATERIALS MODIFICATION

It is generally understood in the plastics molding industry that the materials used, as injection molding materials, may contain small amounts of additives. Such additives may include stabilizers and lubricants. Materials modification usually means the addition of larger amounts of additives. Such additives may include elastomers, flame retardants and fillers.

Materials modification, such as with fibers or with fillers, is extensively adopted with thermosetting plastics, and now with engineering plastics, so as to obtain a desirable combination of properties. It is not done simply to save money as often a molding, made from a filled compound, is the same price as one made from the unfilled plastics material. This is because of the high density of most fillers and because of the high compounding costs. Many of the fillers used are fibrous fillers as the use of such materials improves properties such as modulus.

INJECTION MOLDING OF THERMOPLASTICS

1 COMMON NAMES

Most thermoplastics begin with ' poly' and then this term, which means 'many' is followed by the old fashioned name for the monomer from which the plastic is derived i.e the name is source-based. Because of this practice we get names such as polystyrene and polyethylene for homopolymers. When the plastics material has more than one word in the name, parentheses, or brackets, may be put around the words so that we get poly(vinyl chloride). However this practice is not universal and so we also meet the same term without the brackets i.e. polyvinyl chloride. Source-based nomenclature is not however, universally used and so names such as 'acetals' and 'cellulosics' are also encountered. Other, more logical systems of nomenclature have been proposed but as yet, they have not been adopted,

2 ABBREVIATION(S)

Because of the complex, chemical names used to describe polymers, they are nearly always referred to by abbreviations. Such abbreviations take the form of a short string of capital letters; each capital letter refers to a part of the common name. If the plastics material begins with 'poly' then the first letter is P: the other letter(s) are derived from the monomer unit. Names such as polystyrene and polyethylene are thus shortened to PS and PE respectively (see table 1).

SOME RELEVANT STANDARDS

Many standards organizations, for example, the British Standards Institution (BSI), the International Standards Organisation (ISO) and the American Society for Testing and Materials (ASTM) issue standards which specify what letters shall be used. See ASTM standard D1600-86 (that is standard number D1600 revised/published in 1986) and called 'Standard abbreviations of terms related to plastics. There was a British standard published in 1978 - see BS 3502 1978: the corresponding ISO standard is ISO 1043 1978. There is now another ISO standard ISO 1043-1 which was published in 1987 and which is called 'Basic polymers and their special characteristics'. It has the number ISO 1043-1 : 1987 (E). (The nomenclature of rubbers and lattices are described in ISO 1629).

STANDARD AND-NON STANDARD ABBREVIATIONS

It must be appreciated that both standard and non-standard abbreviations are used. Often materials appear, and become used, before the standards committee has issued its suggestion. By this time a 'non standard' abbreviation has become established or more than one abbreviation may 'appear' for the same material. The thermoplastic elastomer, known as polyether ester elastomer, may be referred to as PEEL or, as COPE (from copolyester) or, as TEEE (thermoplastic elastomer ether ester) or, as YBPO (an American suggestion).

GENERALLY ACCEPTED RULES

Because of the possibility of confusion it is suggested that when abbreviations are used in publications that their first occurrence in the text should be enclosed in parentheses i.e. brackets, and preceeded by the written word being abbreviated. So, for the first time, polystyrene would appear in an article as polystyrene (PS). Subsequent references to such words in the publication can then be by means of the abbreviation.

One practice which does seem to be growing, particularly within companies, is that of increasing the number of letters used to identify plastics, to a minimum of three; this is apparently to suit the needs of common electronic data processing (EDP) systems. If asked to abstract information from a data base, then the request can be made more specific if, for example, polypropylene is given an extra letter, for example R or N so as to make the terminology PPR or PPN, rather than PP. With some systems their is a danger of pulling out everything which begins with PP (for example, PPE and PPO as well as PP) if only PP is asked for. However, this practice should be used with caution as R is used to denote a random copolymer so, strictly speaking, PPR means a random copolymer of PP - these materials are being used in blow molding. Polystyrene may be referred to as general purpose polystyrene GPPS or as PS-GP.

Another practice which seems to be growing, again to suit EDP, is that of not putting anything before the abbreviation for the plastic; for example in this system, low density polyethylene would be referred to as PE LD and not LDPE. In practice, a hyphen is inserted between the groups of letters so that PE LD becomes PE-LD.

ABBREVIATION MODIFICATION

By standard, and by general usage, additional letters may be used to modify those used as abbreviations (see table 2). For example, PE-UHMW therefore stands for ultra high molecular weight polyethylene. As ASTM 1600-86 suggests that GP stands for 'general purpose', PS-GP stands for general purpose polystyrene. In the thermoplastics world, G stands for glass and F stands for fibre/fiber i.e. GF stands for glass fiber.

What a particular letter stands for sometimes depends on the context as individual letters can have more than one meaning, for example, M: when used in GMT it stands for 'mat' and so, GMT stands for a 'glass mat thermoplastics' material. Because of the growth of oriented plastics products, the letter O is put in front the abbreviation for a particular material so, oriented PVC becomes OPVC: this means that a container is made from that oriented material.

Please note that if a letter is not being used in table 2b then, it does not mean that it is not being used in another branch of the plastics industry. (See also the draft ISO standards which are referred to as ISO/DIS 1043/2 and ISO/DIS 1043/3).

POLYAMIDES

By standard, and by general usage, PA stands for 'polyamide'. Numbers after the PA, designate plastics prepared from various condensation units in an homologous series. Of the large number of polyamides possible, two are of great commercial significance; these are PA 6 (polymer of ε-caprolactam) and PA 66 (polymer of hexamethylenediamine and adipic acid). By an extension of the abbreviation system, PA 66, GF 35 means that the material being referred to is 'polyamide 66 with 35%, by weight, of glass fiber'.

COPOLYMERS

If two monomers are used to make a plastics material then, the resultant product is known as a copolymer. It may be referred to by means of initials which represent the monomers used i.e. without a 'p' for 'poly'. For example, the copolymer (bipolymer) made from styrene and acrylonitrile, is referred to as styrene acrylonitrile copolymer or, as SAN. It is now suggested that an oblique stroke / be placed between the two monomer abbreviations e.g. E/P for an ethylene propylene copolymer. However, this suggestion is not universally adopted as SAN is still SAN: the oblique strokes may be omitted when common usage so dictates according to ISO 1043-1 : 1987 (E).

The major ingredient/monomer is usually mentioned first and the other polymer is often only mentioned if it is above a certain percentage, for example, 5%. (Some, so-called homopolymers are in actual fact copolymers but the second monomer is only present in minor amounts).

BLENDS OR ALLOYS

Because of their ease of manufacture, using (for example, twin-screw compounding extruders) there is a lot of interest in blends of plastics or, in blends of plastics with elastomers: either may be modified with fillers or glass fibre/fiber. When mixtures are made from two or more polymers (blends or alloys), it is further suggested that the symbols for the starting polymers be separated by a plus sign, and the whole be placed in parentheses i.e. (A + B). If styrene acrylonitrile copolymer or, SAN is blended with the copolymer made from ethylene and vinyl acetate or, E/VAC then it would be represented as (SAN + E/VAC) in the 'standard' system i.e by ISO 1043-1 1987 (E). More commonly however, it would be referred to as SAN/EVA or, as SAN/EVAC. The major ingredient is usually mentioned first and the other polymer is often only mentioned if it is above a certain percentage, for example, 5%. If below this figure it is regarded as an additive and these are are not often mentioned in commonly used systems.

In the case of both copolymers and blends, it would seem reasonable to indicate the percentage by weight of each ingredient, for example, 70:30, but this information is not often not, readily available.

MORE COMPLEX CODING

A data block system is available, for example in ASTM, to classify, or codify, plastics materials so that, the material description says if the material is, for example, a copolymer based on specified monomers, the intended processing method, whether the material contains additives which modify its basic behavior and, information on its properties. However, many users do not appreciate that such information exists and may be obtained from raw material suppliers.

ABBREVIATIONS USED FOR VULCANIZED ELASTOMERS

Rubbery materials, and compositions based upon them, are commonly referred to by a number of letters which refer to the monomers on which the polymer is based. The standard recommended practice (D1418-72A) issued by the American Society for Testing Materials (ASTM) is the most widely used for the nomenclature of elastomers (rubbers) and lattices. This practice recommends that the elastomers be grouped and coded into a number of classes according to the chemical composition of the polymer chain. Of the various types, the "R" and "M" classes are the most commercially important.

The "M" class includes:-
  CM - Chloro-polyethylene.
  EPDM - Terpolymer of ethylene, propylene and a diene with the residual unsaturated portion of the diene in the side chain.
  EPM - Copolymers of ethylene and propylene.

The "R" class is defined by inserting the name of the monomer(s) before the word "rubber" from which it was prepared (except for natural rubber). The letter immediately preceding the letter R signifies the diolefin from which the rubber (elastomer) was prepared - except for natural rubber. Any letter(s) preceding the diolefin letter signifies the comonomer(s). Commonly encountered members of this class are:-
  BR - butadiene rubber (BR is also known as polybutadiene);
  CR - chloroprene rubber (CR is also known as polychloroprene);
  IIR - isobutene-isoprene.
  IR - isoprene synthetic (IR is also known as cis-polyisoprene);
  NBR - nitrile-butadiene (NBR is also known as acrylonitrile butadiene rubber or, as nitrile butadiene rubber);
  NR - natural rubber; and,
  SBR - styrene-butadiene rubber.

ABBREVIATIONS FOR THERMOPLASTIC ELASTOMERS

The rubbery materials, which are processed as thermoplastics, are usually treated as thermoplastics and identified in the same way. However, they are sometimes referred as TPE (which stands for thermoplastic elastomer) and then, these letters are followed by another letter which identifies the type of TPE. For example, TPE-A means that it is amide based, TPE-U means that is urethane based and, TPE-O means that it is olefin based.

3 ALTERNATIVE NAMES

Many plastics are known by more than one name, for example, acetals may be known as polyformaldehyde or, polyoxymethylene. When this is so then the alternative name(s) should be listed so that cross-referencing is possible. The use of trade names is also prevalent - some of these are shown in Table 3.

4 SOME SUPPLIERS

These are not necessarily the manufacturers as like any other commodity, plastics are bought and sold on a world-wide basis: some companies also specialize in making plastics compounds by, for example, adding fillers and in such a case, the resultant compound may be marketed under their own trade name. See table 3 for a list of some abbreviations and trade names/trade marks.

5 TRADE NAMES OR TRADE MARKS

Where possible the name associated with a particular Company's polymer should be known- this is because in some molding shops, plastics are often only known by their trade name or trade mark. An alphabetical list of some commonly encountered trade names/trade marks, their abbreviations and suppliers, is given in table 4.

6 MATERIAL PROPERTIES

In this section an appreciation of the properties of this particular type of plastic will be given. However, because each of the plastics discussed is, in effect, a family of materials which differ in, for example, molecular weight and molecular weight distribution, then the advice should only be regarded as a general guideline. The properties of plastics may also be dramatically changed by the processing conditions employed and by the use of additives.

PLASTICS AND POLYMERS

Most commercial plastics, also known as resins in North America, are based on the element carbon and are synthesized, or made, from simple, oil-based raw materials. These starting materials are called monomers and these simple, low molecular weight materials are put together, by a process known as polymerization (polymerisation), so as to form polymers. This term means that the final product consists of many identical, repeat units. Because the final molecular weight, or mass, is so large the material may also be referred to as a 'high polymer' or, as a 'macromolecule'.

All plastics are polymers but not all polymers are plastics. Cellulose is a polymer but it cannot be processed like a plastics material unless it is modified.

A plastic is a polymer, which is capable of being shaped or molded under conditions of moderate temperature and pressure. There are two main categories of plastic and these are thermoplastics and thermosetting plastics (thermosets). Thermoplastic products, for example an injection molding, may be softened and reshaped whereas a thermoset product cannot. In terms of tonnage thermoplastics are by far and away the most important. Many plastics are based on one monomer and are known as 'homopolymers': some are based on two monomers and are known as 'copolymers'.

AMORPHOUS AND CRYSTALLINE

Thermoplastics materials may be divided into two main categories; these are amorphous and crystalline. An amophous, thermoplastics material is usually a hard, clear, rigid material with a low shrinkage and a low impact strength. A crystalline plastic also contains amorphous material and so may also be known as a semi-crystalline, thermoplastics material. Such plastics are usually tougher, softer, but can have a higher heat distortion temperature, than an amorphous, thermoplastics material: such plastics are also translucent, or opaque, have a high shrinkage and a high specific heat.

LONG CHAIN STRUCTURE

The 'high polymer molecules', or 'macromolecules', on which plastics are based are incredibly long and, because of the carbon to carbon bond, on which they are based, they are also not normally straight: they are twisted or coiled. (There may be 50,000 atoms linked together in a long chain-like molecule or structure: the chains are also of different lengths.) When there is no order in the system, i.e. if it is an amorphous material, the chains prefer to exist in a random, coiled-up state (just like the shape you would get if you dropped a piece of string on the floor). The separate chains (the separate, long chain molecules) are also normally intertwined one with the other. If crystallization is possible then the chains, or portions of the chains, may lie side-by-side in a more ordered fashion. Because of, for example, the length of the chains involved and the chain entanglements involved, a thermoplastics material does not crystallize completely on cooling; this is why it is called 'a semi-crystalline, thermoplastics material'. Such crystalline materials contain both crystalline and amorphous regions.

ORIENTATION

When a molten polymer is made to flow then this flow occurs because the polymer chains slip, or slide, one over the other. Layers slip one over the other in what is called 'laminar flow'. As the individual, plastics molecules move, one relative to the other, then this causes the molecules to change their direction or 'orientation'. (Reasons for this change are, for example, chain entanglements and friction between layers). The chains become 'drawn out' in the direction of flow and, because of the rapid cooling employed in injection and blow molding, this orientation is 'frozen in': the product contains what is referred to as, frozen in stresses or, frozen in strains. The moldings contains molecules which are orientated in the flow direction: this means that we have a grain effect just like the one that exists in wood. Because of the grain, the wood is stronger in one direction that it is in another.

EFFECT OF ORIENTATION ON PROPERTIES

Because of the grain effect in wood, the wood is much stronger in one direction that it is in another. Injection moldings are also much stronger in one direction than they are in another. However, it is not often that commercial use can be made of this effect in injection molding. This is because if an injection molding is struck or dropped then, it always fails in the weakest direction. This is in the flow direction; it fails 'across the grain'. This is because of the weak interchain forces which bind the orientated molecules together. If the molding could be biaxial orientated then, useful products could be made.

ADDITIVES

By the use of additives it is possible to get a wide range of properties for each material group. Any properties quoted in the literature should therefore only be used as a general guideline. (The properties of plastics may also be dramatically changed by the processing conditions employed.) With many materials, variations of basic formulas are available with additives to provide, for example, improved heat resistance or, weatherability. Some formulations offer improved impact strength or, ease of flow: others, which contain fillers, are used where the moldings require greater impact strength, tensile strength and heat distortion temperature. (Materials modification, such as with fibers or with fillers, is extensively adopted with engineering plastics so as to obtain a desirable combination of properties).

Processing and performance modifiers can also be added; these include antistatic, nucleating and mold release agents. Concentrates containing foaming agents are also available for many materials so that structural foam molding may be performed. Such additives may form part of the color masterbatch.

FEED FORM

The molding machine can be fed with plastics (resins) or compounds in various forms: the feed may be fine powder, regranulated material or, pellets. If the material is available in more than one feed form, then feeding problems will probably be encountered if a mixture of feed forms is used. In terms of feeding efficiency, spherical granules (of approximately 3mm (0.118in) diameter) are the most efficient. Fine powder is usually the worst, followed by regranulated material, then by cube cut granules and then by, lace cut granules (granules made by chopping strands with a circular cross-section). For many good and valid reasons, molding machines are often fed with a mixture of plastics material and masterbatch. The use of such a mixture can however, cause problems. The usual problem is one of color shade differences between different machines; another is separation of the masterbatch from the plastic in the hopper.

7 EASE OF FLOW

Plastics materials differ widely in their viscosity, or ease of flow, and the problem is made more difficult by the fact that each material is available in a range of grades each of which also has a different flow behavior. The position is made even more complicated by the fact that the flow properties of plastics are non-Newtonian and so there is not a linear relationship between pressure and flow. What all this means is that the flow properties cannot be represented meaningfully by one figure and so flow testing over a range of conditions is required.

Such information is available in the materials data base maintained by Moldflow (Europe) Ltd. and where possible some of that information will be quoted so as to illustrate the flow properties possible. The effect of changing temperature at a constant shear rate will be given together with values which show the effect of changing shear rate at a constant temperature. Shear rate will be given in reciprocal seconds (that is, in s^{-1}) and the larger the number quoted the faster is the material being sheared or forced along. In most cases, plastics are pseudoplastic and all this means is that they become less viscous (more easier flowing) if they are moved quickly. Look at the table in the 'Ease of flow' section for a material, for example, PA6 (nylon 6).

MOLDFLOW DATA FOR PA6.

Grade. Akzo 'Akulon'	Viscosity (Nsm^{-2}) at $1,000s^{-1}$ at the following temperatures			
	°C/°F 240/464	°C/°F 260/500	°C/°F 280/536	°C/°F 300/572
Easy flow grade	98	78	63	50
Medium flow grade	149	115	89	69
Stiff flow grade	394	319	259	210

Grade.	Viscosity (Nsm^{-2}) at 280°C (536°F)			
	$100s^{-1}$	$1,000s^{-1}$	$10,000s^{-1}$	$100,000s^{-1}$
Easy flow grade	197	63	20	6
Medium flow grade	323	89	25	7
Stiff flow grade	1220	259	55	12

These figures clearly show that the three grades quoted have very different viscosities, with the first grade having the lowest viscosity that is, it is the easiest flowing grade (the lower the number, the lower is the viscosity). It would be used where mold filling is difficult, or where long flow lengths are involved. The viscosity of all three materials falls as the shear rate is increased, that is, mold filling becomes easier. Raising the melt temperature reduces the amount of injection pressure required to maintain a certain rate of flow.

The viscosity of most plastics falls as the temperature is raised but, with some materials the effect is more pronounced. So, for materials whose viscosity falls markedly with a rise in temperature (for example, PA66, PC and POM) ease of mold filling is really helped by a moderate increase in melt temperatures. For other materials, where the effect of temperature increases is less marked, it is more useful to increase the mold temperatures and/or, the molding pressures, to ease mold filling.

Once a plastic has been selected for a particular application, then a high viscosity grade should be selected if the components are to be subjected to severe mechanical stresses; this is because the high viscosity grades usually have the highest molecular weight and exhibit the best mechanical properties. However, in some cases this advice cannot be followed as unacceptable levels of frozen-in strains result. Easy flow grades are preferred for filling thin walled sections or, for use where very smooth surfaces are specified.

8 SHRINKAGE

Most plastics components are smaller than the mold used to produce them. The magnitude of the shrinkage differs from an amorphous, thermoplastic material to a semi-crystalline, thermoplastics material by a factor of approximately four. Shrinkage may also be different from one grade of material to another and will certainly be influenced by changing the processing conditions; it may also be significantly different in different directions, for example, across the flow and along the flow. For a semi-crystalline, thermoplastics material the shrinkage in the direction of flow is approximately twice the shrinkage found across the flow direction.

Shrinkage may also be different at different points along the same flow direction, for example, with increasing flow path lengths, the mold cavity pressure drops and this will lead to higher shrinkages in these low pressure areas. With increasing wall thickness, shrinkage also increases so the shrinkage can vary in this dimension (this variation can lead to distortion if the molding contains wall thickness variations). Because of these, and other, factors it is usual to quote a shrinkage range for each plastics material, for example, 0.004 to 0.008in/in or, 0.004 to 0.008mm/mm.

The total shrinkage (TS), experienced by a molding, is made up of mold shrinkage (MS) and, post-molding shrinkage or after shrinkage (AS). Mold shrinkage is defined as the change in dimensions between the size of the cavity and the molding, 24 hours after the molding is ejected from the mold. (Mold shrinkage is quoted as a percentage, for example, 0.4%, or as a linear shrinkage, for example, 0.004in/in or 0.004mm/mm.) Post-molding shrinkage occurs after the molding has been aged and is also called environmental shrinkage.

Usually mold shrinkage increases when increasing mold and/or melt temperature. However, total shrinkage often decreases as mold temperature is increased and, in practice, more dimensionally stable moldings are obtained. If the mold temperature is very high, then post-molding shrinkage may be neglected even for semi-crystalline, thermoplastics materials. Such high mold temperatures are used where close tolerances are required.

Because of the large, and variable, shrinkage associated with plastics materials, it is unreasonable to expect to obtain the tolerances possible when metals are machined. To assist designers, various bodies (e.g. DIN) issue advice on the tolerances possible when specific materials are molded. For example, materials may be divided into three groups:-

(i) contains those plastics materials with a shrinkage below 1%, for example, thermosets and amorphous, thermoplastics materials;

(ii) contains those plastics materials with a shrinkage between 1 and 2%, for example, PA, POM, PBT and, PET; and,

(iii) contains those plastics materials with a shrinkage between 2 and 3%, for example, HDPE and PP.

The materials in the first group are capable of being molded to the tightest tolerances: the tolerance band will increase as the component size increases. A material may be moved from one group to another by, for example, filling. Adding filler to a PA, can move it from group one to group two.

9 RESISTANT TO THE FOLLOWING

When a plastics material is stated as being resistant to a certain chemical then this should only be taken as a guideline as there are usually many different grades of each material and some of them can be very different. Nevertheless, it can be said that plastics have, in general, good chemical resistance. The chemical resistance is again, in general, better for a semi-crystalline, thermoplastics material than for an amorphous, thermoplastic material. With all plastics, chemical resistance worsens as the temperature is raised. When checking on the properties of injection molded plastics it must not be forgotten that processing can dramatically affect the properties of plastics. If it is known that an injection molding will be used in a particular environment then it is worthwhile checking to see if the materials manufacturer has any relevant data as sometimes catastrophic failure occurs, in what seems to be, a relatively harmless environment.

The chemical resistance of a plastics material is therefore very dependent upon the particular application and closely linked with the time and temperature of exposure. Where extremes of environment may be encountered or, combinations of chemicals met, then additional information is required before that material may be specified.

Advice in this section is usually given on the chemical resistance to strong and weak acids, to alkalis and to solvents. A strong acid, or a strong alkali, remains 'strong' by definition, even when it is used diluted: a weak acid, by the same token, is still 'weak' even when it is used concentrated. Solvents may be polar or non-polar. The term means that the material is not balanced and so charge separation on a molecular scale exists. Examples of polar solvents include water, alcohols, ketones, esters, and partially halogenated solvents. Non-polar solvents include carbon tetrachloride, benzene and hydrocarbons.

10 NOT RESISTANT TO

The short term, chemical resistance of most plastics is well known. What is not so well known is the long term chemical resistance or the environmental stress cracking resistance (ESC).

At first glance a plastic may appear to resist a particular chemical, for example water; however a long term study, for example over hundreds of hours, may reveal a different story. Testing of actual moldings, in a particular environment, is the only answer. The problem of environmental stress cracking can be even more difficult to solve in some cases. The term means that if a molding is stressed in a particular environment, for example a chemical which may appear harmless, then it will fail. This eventuality should always be considered.

The difference in behavior between plastics and metals should also be remembered. With metals, and other materials, chemical attack is usually confined to the surface and involves corrosion and a loss of weight. With plastics however, there is often absorption of the chemical reagent, leading to swelling and softening, followed by a gain in weight. In extreme cases, this may be followed by a loss of weight as the surface is dissolved or, the surface becomes so swollen, or decomposed, that it crumbles away.

11 MATERIAL DETECTION OR IDENTIFICATION

Only those tests which can be performed easily and cheaply will be described. These include density, melting point and behavior on heating. The tests should be done in the order shown; if the density determination is high, for example greater than 1.7 gcm^{-3}, (SG 1.7) the material could well be a fluoropolymer and, heating tests should not be done because of the danger from the fumes evolved. In any heating test, only a very small sample should be employed for the same reason; ideally a fume cupboard should be used.

12 COLORING

Coloring is extremely important to the plastics industry as, after all, one of the major reasons for the use of plastics materials is the bright attractive colors possible.

COMPOUNDED COLOR

Amorphous thermoplastic materials are transparent and so are available in a wider color range than semi-crystalline, thermoplastics materials. Traditionally all plastics were molded from colored granules but now, because of the ease with which color may be added at the machine, there is increasing use in the injection molding industry of color concentrates, or masterbatches, in conjunction with natural, or un-colored, material.

However, compounding still remains the most accurate coloring technique and gives the most precise, and reproducible, color. It also gives the best density of color and is the most suitable for small runs. Most engineering plastics (EPs) are sold already colored at the present time but it is thought that, despite color uniformity problems from machine-to-machine, coloring on the injection molding machine will increase; particularly by the use of solid masterbatches.

MASTERBATCH COLORING

Masterbatches may be solid or liquid and an extremely wide color range is possible from such blends. The solid ones are the most popular and may be based on a wax-like or resin based, carrier (a universal carrier) or, on the base polymer (polymer specific concentrate). The latter are the cheaper of the two. Before a universal masterbatch is used it is important to check that it is suitable for use with the plastic; some masterbatches are claimed to be universal but this claim should be treated with caution. Differences in shade will probably be seen when changing from one plastic to another even though the masterbatch concentration is kept constant. Differences in shade may often be seen even if the same plastic/masterbatch combination is being run on different machines. It should not be forgotten that because of the high pigment concentration (e.g 50%) the density of most masterbatches is very high; far higher that the plastic they are used with. This density difference can cause problems of separation, particularly if batch mixing is employed.

Despite initial optimism, liquid colors are now not widely used. Their most successful applications seem to be on long runs with one color where the precise metering for uniform color can be established and held. A major problem with liquid color is one of cleaning up if they are spilt or, not dosed correctly.

DRY COLOR

The cheapest way of coloring plastics is by the use of dry colorants, however one main disadvantage is that it tends to be a messy and dusty process. To ensure color consistency throughout a production run (i.e. to improve accuracy and consistency), sachets or cartons possessing the correct amount are available for adding to standard sized bags (for example, 25kg/55.1lb) or, to other, larger quantities. When using dry colorants it is imperative that the surface of the granules are uniformly covered with the colorant in order to achieve satisfactory dispersion of colorant in the polymer melt. The type of blending technique (e.g.

single or multi-axis rotation) and blending time should be standardized, so as to ensure that uniform coverage of granules by colorant is obtained. Once a successful procedure has been established this should be maintained in order to achieve color uniformity. Care must also be taken to ensure that the colorant does not absorb moisture during storage as agglomerates will readily occur: such agglomerates can give color shading or streaking in the molding. Because of toxicity problems many powdered pigments (e.g. cadmium based) require careful handling: such problems help explain why the use of dry colors is declining.

Whatever the method used to color the natural polymer, it must be borne in mind that the properties and dimensions of the molded part can be considerably affected by the type of dye or pigment used.

SCREW DESIGN AND COLORING

Screw design is one of the most important factors in determining the over-all performance of an injection molding machine. However, it is not always given the consideration it deserves, particularly with the increase in coloring now done on the machine by, for example, solid masterbatches. A traditional screw is often not very efficient at this relatively simple mixing operation even when high melt temperatures and high back pressures are used. A two-stage, vented barrel, screw will usually give better results: screws equipped with mixing sections should also be considered.

13 MATERIALS AND COMPONENTS HANDLING

Far too often a lot of time and money is spent on improving the injection molding operation while ignoring this very important subject.

MATERIAL SUPPLY AND HANDLING

Most materials are supplied as granules or pellets and are usually supplied in either sacks or, in bulk containers. When supplied in 50lb/22.6kg sacks, then the material is often supplied in 20 sack (1,000lb) loads. If supplied in 25kg/55.1lb sacks then, the material is often supplied in 1,000kg/2240lb lots.

Supplying the material in sacks, or bags, is not the only way that the material can be supplied: considerable discounts may be obtained if the plastics material is purchased in bulk and, supplied in, for example, bulk containers. With some materials, this form of supply may not be acceptable as, for example, excessive water absorption may occur.

No matter how they are delivered, the materials must be carefully looked after as plastics are expensive and many will burn relatively easily. Plastics materials are contaminated extremely easily simply by leaving sacks, containers, or hoppers, open. Therefore always keep materials covered.

All storage and unloading areas must be kept clean and dry; to minimize the fire hazard, the store rooms should be separated from the injection molding shop by fire resistant doors. Store the materials away from direct sunlight and on properly constructed racks. Usually the use of unheated storage areas, with natural ventilation, is sufficient. Ensure that the material does not stagnate in the stores by adopting a strict stock control policy. Adopt a strict first-in, first-out policy (Fi-Fo). Strict stock control is important, as within a shipment of material there could be several different batches and one of these could be faulty. By adopting a strict stock control system, a faulty batch can be readily identified and isolated thus preventing further defective moldings being produced.

If the material is available in more than one feed form, then feeding problems will probably be encountered if a mixture of feed forms is used. In terms of output efficiency, spherical granules (of approximately 3.2mm ($1/8$in) diameter) are the most efficient. Fine powder is the worst, followed by regranulated material, then by cube cut granules and then by lace cut granules (granules made by chopping strands with a circular cross-section).

When plastics are conveyed in automatic systems, then the dangers of dust generation should be appreciated. All equipment must be properly earthed.

The subject of materials and component handling is a subject that in many molding shops appears to be ignored so that material and molding contamination is common. The most common cause of material contamination is water and that of moldings is oil and/or grease.

CONTAMINATION AND DRYING

Most plastics materials are supplied dry and ready for use although some are 'wet' and must be dried before being injection molded. Many plastics, and particularly engineering plastics, absorb water (they are hygroscopic) and, if this is so, they must be dried before use (in many shops the materials are dried as a matter of course as this saves any argument). However, it still makes sense to minimize water absorption.

PREVENTING CONDENSATION

When the plastic is brought into the molding shop, allow the material to reach work-shop temperature as this will save the material becoming contaminated by condensation. For example, before the containers are opened, after being brought from store, allow them to stand in the injection molding shop for 8 hours: this applies particularly during the colder months, and at times of high humidity. At such times it is advisable to store the day's requirements in the molding shop.

REDUCING MOISTURE CONTENT

A measure of how much water a material will absorb is given by how much it will absorb in 24 hours at room temperature. If greater than approximately 0.2%, drying is usually necessary; if less, this means that drying is not normally necessary.

The water (moisture) content in a wet material may be reduced to an acceptable level by the use of a vented machine: this will reduce not only the water content but will remove other substances which are gases (volatiles) at injection molding temperatures and will also give very efficient color compounding.

$$\text{where } Q = \frac{\text{total shot weight (g)} \times 3600}{1000 \times \text{cycle time (s)}}$$

$$Q = \frac{96 \times 3600}{1000 \times 24} = 14.4 \text{ kg/h}.$$

or, in pounds per hour;

$$Q = \frac{\text{total shot weight (oz)} \times 3600}{16 \times \text{cycle time (s)}}$$

However, with some materials, for example PC, the use of a vented machine may result in a loss of properties as a result of heating the plastics material while it is in contact with water. Such a loss of properties may not be immediately apparent as the moldings may appear satisfactory; the impact strength in service may not, however, be satisfactory. Because of these reasons, oven drying is often preferred.

If oven drying is required, then dry in either a hot air oven, in a desiccant dryer or, in a vacuum dryer. The last two methods are the more efficient as they are quicker and reduce the water content to lower values. They take approximately half the time necessary when using a hot air drier and, for some materials, they are the only way of getting them dry.

CALCULATING HOPPER SIZE

After drying the material it should not be placed into an open hopper for extended periods of time where it will reabsorb moisture for, when molding hygroscopic materials, such as ABS and PA66, this can produce moldings having splash or splay marks on their surfaces. Material usage for a given machine/mold combination should be calculated so as to determine how much material should be loaded into the hopper to prevent storage for any length of time, preferably using a heated hopper assembly. Care should be taken to ensure that the hygroscopic materials are in the hopper for no longer than 1 hour.

Material usage for a given machine/mold combination should be calculated to determine how much material should be loaded into the hopper to prevent storage for any length of time, preferably using a heated hopper assembly. Care should be taken to ensure that the hygroscopic materials are in the hopper for no longer than 1 hour.

Consider the example where of a machine which is producing 6 components at a cycle time of 24 seconds (24s). The weight of each component and the feed system is 14 grams (14g) and 12g respectively, therefore the total shot weight is (6 x 14) + 12 = 96g. Given the following formula, the material consumption (Q) in kilos per hour (kg/h) can be calculated as follows:

Therefore the hopper should be filled with 14.4 kg or 31.7lbs, the level noted accordingly, and it should not be topped up beyond this point.

MOLDING CONTAMINATION

Oil and/or grease contamination usually occurs after ejection of the moldings (for example, it drops from the toggles, platen bushes and hydraulic cylinder seals) and should be corrected immediately; it should be kept under control by preventative maintenance. (Using all-electric machines it is now possible to produce moldings under clean room conditions). A common cause of grease contamination is grease leeching from around the ejector pins; this is particularly true for ABS. The mold should be stripped and cleaned thoroughly and then, re-assembled using a white PTFE based lubricant.

Many plastics attract dust at a tremendous rate and this can be extremely difficult, and expensive, to remove. It is far better to stop the problem occurring in the first place by, for example, keeping the moldings covered at all times and handling them as little as possible. When they are handled, or moved, the dangers of scratching the injection moldings should be remembered (they will even scratch each other if rubbed gently together); many injection moldings are damaged in transit because of poor packing. If touched with the bare hands then grease can be transferred to the moldings and this, again, can be very troublesome, for example, in electroplating. Clean, dry gloves should be worn when injection moldings are handled unless it has been proved that this precaution is unnecessary.

14 MOLD AND GATE CONSIDERATIONS

Some aspects of mold design will be different for each material and where appropriate these will be mentioned in this section. However, there are general rules which should be applied as far as possible and these are as follows:-

(i) design the mold as a heat exchanger;
(ii) use a correctly sized feed system;
(iii) cool the injection molding uniformly;
(iv) use adequate mold venting;
(v) use standard mold parts and accessories wherever possible;
(vi) use runnerless type molds where convenient;
(vii) use pretoughened steels;
(viii) use computers to assist mold design;
(ix) make proper allowance for shrinkage;,
(x) establish the critical dimensions;
(xi) design for ease of handling and fixing;
(xii) design the mold to be insulated;
(xiii) consider alternatives to steel and, last but by no means least,
(xiv) design the component to suit the injection molding process.

HEAT EXCHANGER DESIGN

It is essential to design the mold as a heat exchanger because of the importance of mold temperature cooling and control on molding cycle time and, on product properties. If the maximum amount of heat that needs to be removed from each gram of shot {given in Joules (J) or, in British thermal units (Btu)} is known, then this figure, may in turn be used to calculate the amount of fluid (usually water) that needs to be circulated through the mold so as to remove the heat carried in by the plastic. This is done, for cold runner molds, by assuming that all of the heat contained in the components must be removed by the mold cooling system but that, only half of the heat contained in the feed system needs to be removed before ejection takes place. To do this, the specific heat of the plastic must also be known; as this changes with temperature, the average specific heat (over the temperature range from the melt temperature to the mold temperature is used.

Specific heat is quoted in J/kgK or, in Btu/lb°F. The specific heat is multiplied by the temperature difference (ΔT) and, by the weight of the total shot to give the heat content - also known as 'enthalpy'. (For hot runner molds, only the moldings are considered; assume that all of the heat contained in the components must be removed by the cooling system.)

To convert from kJ/kg: to Btu/lb multiply this figure by 0.43; see table 6.

The size and location of the cooling/heating channels are extremely important as these allow a rapid and uniform heat removal during the material solidification stage. Cooling/heating channel diameters of 12mm (0.472in) are recommended so that the flow rate of the cooling/heating medium is suitably high enough to maintain the mold surface temperature to within +/− 1°C (+/−1.8°F). To obtain a constant mold temperature it is recommended to use insulating plates attached to the back of each mold half so as to minimize the heat loss. Such plates also reduce the time period needed to heat-up the mold to the required temperature.

CORRECTLY SIZED FEED SYSTEM

Typically runner sizes range from 3 to 10mm (0.118 to 0.394in), the most common size being 6mm (0.236in). Runners, like sprues, are usually short in length and generous in diameter as this reduces pressure loss and thus, permits the application of adequate follow up pressure. However, if they are made too large then excessively long cycles and large material losses result. If they are made too small then, the mold is incapable of being filled and the large amount of pressure which is lost is transferred into heat: heat generation in injection molding is proportional to the pressure drop in the process. This heat will show up in the regions where the material is being sheared the most, i.e. in the gate regions. Such local temperature rises can be very high and can lead to material degradation. (This is why maximum shear rates are sometimes quoted for plastic materials). It is generally true that, in the plastics industry, runners are made too large and gates are made too small.

UNIFORM COOLING

Cooling the injection molding uniformly may mean cooling the mold at different rates, in different areas, so as to get uniformity of component cooling. The aim must be to cool the component as quickly as possible while ensuring that faults such as poor surface appearance, changes in physical properties etc. are not encountered. Each part of the molding should be cooled at the same rate. This often means that non-uniform cooling must be applied to the mold, for example, cool water should be fed into the inner parts of the mold cooling system (particularly in the area of the gate) and, warmer water into the outer parts. This technique is essential when molding flat components to close tolerances, or large components that include long melt flow lengths from the gating position.

ADEQUATE MOLD VENTING

The mold must be vented to allow for gas escape; such vents must be placed near weld lines and also near the last areas to be filled. Typical vents are slots 6 to 13mm (0.236 to 0.512in) wide and 0.01 to 0.03mm (0.0004 to 0.001in) deep; such slots are located on the mating surface of one of the mold halves. If a negative pressure device is available it may be possible to vent the mold into the water channels. This can speed up mold filling, reduce component burning and, reduce the cycle times.

STANDARD COMPONENTS

Use standard mold parts and accessories wherever possible. Many molds, approximately 70%, are now designed around standardized components as this can speed up the drafting process (the standardized components may be held in a computerized system) and, their use can also reduce mold costs – both production costs and, maintenance costs. The range of components now offered by some companies e.g. DME, Hasco, is now very impressive with items such as the following now being standard:

    a) date stamps for moldings - useful for quality control;

    b) temperature controlled sprue bushes - useful for thermoset molding or, for reducing cycle time when molding heavy-sprued components;

    c) positive locking on side cores - this has now been integrated with the side core actuator and so simplifies design and operation;

    d) reverse taper nozzles - these are used for large molds, e.g in PAGF and PC, where there is a need for a central, pin-point gate but a three-plate mold is not wanted. The sprue is ejected pneumatically, towards the injection unit;

    e) rapid mounting jigs or plates - these are used for mold mounting where, for example, the mold is fitted with studs which locate, and are locked into, holes in the plates;

    f) rapid couplings - for example, for water and for the ejector system;

    g) threaded, tapered cartridge heaters - these give better control and ease of heater removal in the event of heater failure;

    h) two stage ejection units - useful where side cores are used; and,

    i) side core movement devices - useful for undercut type molds

USE RUNNERLESS TYPE MOLDS

The use of runnerless molds is becoming more readily accepted by the molding industry due to their high success rate. The variety and types of systems available nowadays are numerous and therefore one should discuss in detail, with the hot runner, and insulated runner specialists, the requirements/quality of product to be manufactured from a particular polymer. This is necessary in order to ascertain which system will be best suited to a particular job. It is a well known fact that detailed investigations carried out at the design stage, concerning runnerless type molds, save considerable production problems. When long sprues and/or runners are being molded, then the use of a runnerless mold can speed up cycle times as the mold opening and closing movements may be shortened considerably: as the sprue and runners are the thickest part of many moldings, their elimination can also save on cycle time by reducing the mold cooling time. It is well worth considering the use of cast heaters for runnerless-type molds as they last so much longer than conventional heaters. In thermoset molding, the use of hollow sprues (hollow cone molding) can result in large cycle time reductions.

PRETOUGHENED STEELS

The majority of injection molds used nowadays are manufactured from nickel-chrome alloyed steels which are subsequently heat, and/or surface, treated in order to obtain a surface hardness ranging from 48 to 67 Rockwell C. Such a hardness is used to prevent damage occurring to the mold surfaces, to obtain a good surface finish and, to increase stability/life usage of the mold. Many moldmakers are now beginning to use pretoughened steels, without further heat treatment, in order to reduce cost and delivery times. One reason for the use of such steels is the improvements and conciseness of control with respect to the mold sensing and closing/locking operation of the molding machine. This has led to less damage occurring on mold faces when moldings, or feed systems, become entrapped between the mold halves. However, some moldmakers are not yet fully convinced that suitable dimensional stability and robustness is achieved for fast cycling, multicavity, production molds, made from pretoughened steel, and will therefore continue to use hardened molds until proven otherwise.

COMPUTERIZED MOLD DESIGN

The standard or quality of moldings is ever increasing and the need to accurately predict the material's flow and cooling characteristics within the mold cavity is becoming a necessity. The advent of computer aided mold design has made this task much easier enabling detailed analyses to be carried out for a range of processing conditions and, for a range of gating positions. The results of such analyses enables molds to be designed and manufactured knowing that a successful component will be produced with the minimum of debugging and that target dates can be met.

SHRINKAGE ALLOWANCES

Most plastics components are smaller than the mold used to produce them. Mold shrinkage is not however, the end of the story as the total shrinkage (TS), experienced by a molding, is made up of mold shrinkage (MS) and post-molding shrinkage or, as it is sometimes known, after shrinkage (AS). Mold shrinkage is defined as the change in dimensions between the size of the cavity and the molding, 24 hours after the molding is ejected from the mold. Post-molding shrinkage occurs after the molding has been aged and is also called environmental shrinkage.

To obtain values for mold shrinkage, components are produced by an appropriate molding method: compression molding for thermosets and injection molding for thermoplastics materials. After a specified time (eg 2h) the dimensions of the molding are measured at room temperature and the dimensions of the cavity, if not known, are also measured. The mold shrinkage (MS) is given as a percentage by $100 \times L_0 - L_1 / L_0$ where L_0 is the length of the cavity and L_1 is the length of the molding. Post-molding shrinkage (AS or PMS) is given as a percentage by:

$$PMS_{48h} = 100 \times L_1 - L_2 / L_1$$

where L_1 is the length of the original molding and L_2 is the length measured after 48h (or some other time). To convert from volumetric shrinkage (MS_v) to linear shrinkage (MS_L) use:

$$MS_v = -1 + (1 + MS_L).$$

When molding shrinkage is used to calculate cavity dimensions then use:-

$$D_c \text{ (cavity dimensions)} = D_p + D_pS + D_pS^2.$$

Where D_p is the dimensions of the molding and S is the linear shrinkage.

Because shrinkage is so dependent upon the production conditions it is usual to quote a range for each material. The actual value used to size a cavity may need to be obtained by producing a pilot cavity and then, producing components under production conditions. (If a molding is produced by a cold runner system, which is then changed to a hot runner system, then it will be probably found that different shrinkages are obtained. If these cannot be compensated for by setting changes then the original design will have to be re-used).

ESTABLISH THE CRITICAL DIMENSIONS

The end user, the component designer and, the injection molder must all agree on the dimensions of the moldings before production commences. The size and number of critical dimensions must be established very early in the design stage and, the tolerance assigned to each dimension must be made as wide as is possible. That is, a dimension should not be specified to within certain tolerance unless, it is essential (critical) that it be held to those sizes.

DESIGN FOR EASE OF HANDLING AND FIXING

Many molds seem to be designed without due regard for ease of handling, mounting or fixing. Unless this is done then a lot of valuable production time will be lost while the mold is being fitted to the injection molding machine i.e. being set up. To save valuable production time it is suggested that:-

(a) all inlet and outlet, cooling connections are clearly stamped/identified on the mold;

(b) all cooling connections are put on one side of the mold and are designed for rapid coupling;

(c) cooling connections are sensibly located so that they are accessible (e.g not located behind tie-bars) and do not hinder mold mounting;

(c) cooling connections are sensibly located/protected so that they are not damaged during storage, transport and use;

(d) cooling hoses are of standard diameters, lengths and fittings;

(e) mold heights are standardized so that a particular mold is of the same height as others of a similar type in a factory;

(f) the mold base plates are of a standard thickness or, the points of mold mounting/fastening are of the same thickness;

(g) lifting and handling features, for example, eyebolts, are incorporated into the mold at the design stage;

(h) take advantage of the rapid mold mounting, systems now available (quick action, coupling systems are now available for both mold mounting and for the heating/cooling systems);

(i) have setting up, instruction sheets available which are written with the setter in mind and which, not only give precise instructions, but which also state what equipment, services and personnel are needed to get that mold into production;

(j) the mold is got ready as far as possible before mold mounting/setting, on the machine, is started, for example, the mold and associated equipment could be checked and made ready, some hose connections could be made, material could be delivered to the machine, the mold could be pre-heated before fixing and setting etc..

DESIGN THE MOLD TO BE INSULATED

Many molds, particularly those used for engineering thermoplastics, run at relatively high temperatures, for example, 80°C (176°F). If the mold is not insulated then, heat losses to the atmosphere, and to the machine, can easily equal those lost by the injection cylinder. So, insulate the mold from the platens and, if possible, insulate the outside surfaces of the mold. If a hot runner mold is being considered then, try and reduce heat exchanges between the hot runner parts and the cold, component-forming parts.

CONSIDER ALTERNATIVES TO STEEL

Most injection molds are made from tool steel, even though the length of run may not justify this difficult to machine material. Many alternative materials and manufacturing methods are suitable for prototype, pre-production or small batch work and their use can lead to worthwhile cost savings in many cases.

DESIGN THE COMPONENT TO SUIT THE PROCESS

The component must be designed to suit the injection molding process if the best results, in terms of consistency and speed of production, are to obtained. Long trouble-free runs are the ideal and these can only be obtained if the material, the machine and the injection mold are all correct and suited one to the other. The component must be designed, and the mold must be made, so that injection moldings are produced with the minimum of difficulty at the required rate.. Components must eject easily, be free of flash and possess the desired dimensional tolerances. In general, tolerances must be as wide as possible, wall thicknesses must be as thin and as uniform as possible, side walls must be generously tapered and corners rounded. To minimize warping, surfaces must be grooved, corrugated or curved; to ease production, shapes which demand the use of side cores should be avoided as far as is practical.

It should be possible to design, and to operate, injection molds so that high injection molding pressures are not necessary. This may be done by using a correctly sized feed system, using high melt and mold temperatures and, by using reduced flow path: wall thickness ratios. If the injection pressure requirements are reduced then, better components may often be produced more cheaply. Better components because, for example, the amount of shear induced brittleness, in some semi-crystalline, thermoplastics materials, is reduced. The moldings may be cheaper as clamping pressure requirements will be reduced: this in turn will reduce the wear and tear on the machine and on the mold.

15 FLOW PATH : WALL THICKNESS RATIO

This is another way of indicating the ease of flow of a plastics material. If the ratio is quoted as being 150:1 then this means that if the wall thickness of the molding is 1mm, or 0.039", then the maximum length of flow from the gate will be approximately 152mm or 6in. Because the amount of flow possible is dependent upon wall thickness, the flow ratios may be quoted for a range of wall thicknesses.

As the ratio increases for a given thickness, then the amount of pressure needed to fill the cavity will also increase - simply because we want the material to flow further. Making the material flow further, usually means less time for the relaxation of flow-induced orientation and so, the molding may contain large amounts of anisotropy: they will be very much weaker across the flow direction. Components which require large values for the injection molding pressure, for their production, will probably be very anisotropic. (For example, for PA66, it is suggested that the injection pressure should not exceed 8,000psi/56MNm^{-2} so that, excessive anisotropy is avoided.) This means designing the mold to reduce the flow path; wall thickness ratio and, operating the injection molding process so as to minimize anisotropy, for example, by using high mold temperatures.

16 PROJECTED AREA

The projected area of a molding is the largest area that can be seen if the molding is viewed from the direction of clamp force application, For most molding jobs it is approximately 2 tons per square inch (2tsi) or 31 megaNewtons per square meter (31MNm^{-2}). This is however, a low figure and should only be treated as a very rough rule of thumb because, for example, once the molding has any depth then side wall forces must also be considered. If components are being produced by a foam molding process then, the amount of clamping force can be very much lower than that used for the production of solid moldings.

17 CYLINDER EQUIPMENT

It is assumed that the type of machine being used is an in-line screw type; this is because that type of machine is the most popular. Unless otherwise specified such machines are sold with general purpose screws fitted. Such general purpose screws typically have a compression ratio of approximately 2.0/2.5:1 and a length to diameter ratio (L:D) of about 20;1. The root diameter of the screw increases gradually from the end of the feed section to the start of the metering section; the feed section is approximately 50% of the length (50%L), the transition section is 30%L and the metering section is 20%L (including the sliding ring valve).

As general purpose screws are designed to suit as wide a range of plastics as possible it can be said that a general purpose screw will not be the ideal answer for the injection molding of a specific material, for example, a PA. A screw designed for this semi-crystalline, thermoplastics material must provide a greater heat input than one designed for an amorphous, thermoplastic material. If a machine is therefore dedicated to one plastic for a long time then it is well worth considering purchasing a specially designed screw.

No matter what type of screw is used, it is usual to employ a back-flow valve on the tip of the screw and to prevent drooling from the nozzle it is usual to employ decompression (suck-back) or to fit a valve on the machine nozzle.

When back-flow valves are used they must be inspected, and serviced, regularly as they play an important role within the injection unit. These comments apply particularly to high speed machines where screw cushion may not be used and extreme accuracy is demanded. Shut off nozzles are not widely used nowadays, due to material leakage and degradation taking place within the nozzle assembly. Guidance will be given on the types of nozzles used for each material. No matter what nozzle is used, the correct nozzle contact force should be used as, if it is too much, then the fixed mold half may be pushed off the platen.

When abrasive compounds are being processed, then the purchase of a bimetallic, injection cylinder, and a wear-resistant screw, should be seriously considered. If the screw is deep nitrided to 67 Rockwell C this will give protection against chemical attack and also stop plastic adhering, and then decomposing, on the screw.

18 SCREW CUSHION

For most molding runs the amount of screw rotation must be adjusted so that there is always a pad ,or cushion, of material left after the screw has finished injection; this ensures that the screw forward time is effective and a constant injection pressure is being applied. On small injection molding machines this cushion may be 3mm (0.118in); on larger machines it may be 9mm (0.354in); it is usually approximately 10% of the shot volume. Such a screw cushion is necessary on most injection molding jobs, as it transmits the injection pressure from the hydraulic system into the mold: if this transfer did not occur, then the molding would contain voids. To ensure component consistency, the screw cushion on some machines may be held to within +/–0.1mm/0.004in.

On some high speed molding machines, producing say thin walled containers from TPS, there may well be no screw cushion employed. It is dispensed with in order to reduce clamping force requirements and to reduce the risk of flashing. Extreme accuracy is demanded from all components of the injection units of such machines.

19 SHOT CAPACITY

Machines are often rated in terms of the amount of PS that they can uniformly plasticize, or raise to a uniform molding temperature, in one hour. Now PS is an amorphous, thermoplastic material and such materials have a different heat input requirement to that required by a semi-crystalline, thermoplastics material. Because of this, it is usually necessary to under-rate the shot capacity of the injection cylinder when molding materials other than styrene-type plastics. A useful rule of thumb is to use 75 to 80% of the rated barrel capacity so as to minimize melt residence time and to allow injection molding at higher melt temperatures; this will give better melt flow while avoiding decomposition. However, if an injection cylinder is under-utilized (below 35%), then lower melt temperatures will be required if decomposition is to be avoided; this in turn can mean higher injection pressures, more frozen-in stresses and less stable injection moldings.

To determine whether the standard of product quality can be maintained throughout the production run, the total weight of plastics material being injection molded per hour should be compared with the rated plasticizing capacity of the machine. If it is greater than the plasticization capacity then, the molding machine is unable to sufficiently plasticize the material so as, to achieve a uniform melt quality, for example, a uniform viscosity. Therefore the molded components will be subject to some variation (e.g. dimensional or visual). It is particularly important to match shot capacity and plasticizing capacity in the case of fast cycling machines producing thin walled, or close tolerance, components.

To check if the plasticizing capacity, in kg/h or lb/h, is being exceeded, then the total weight molded per hour must first be calculated. Weigh a total shot, that is the weight of both the components and the feed system – call this W (g or oz). Time how long it takes to produce this shot – call this t (which is the cycle time in seconds or s)). To get the number of shots per minute, when the cycle time is <1 minute, divide 60 by t. To get the number of number of shots per hour, multiply this figure by 60. Multiply by

W and convert to either kg/h or lb/h: compare with the plasticizing capacity. Put another way,

$$t = \frac{\text{the total shots molded/hour (g)} \times 3600}{\text{the plasticizing capacity of press (kg/h)} \times 1000}.$$

Or,

$$t = \frac{\text{the total shot molded/hour (oz)} \times 3600}{\text{the plasticizing capacity of press (lb/h)} \times 16}.$$

What this means is that if the mold is run at a cycle time lower than t then, the molding machine is unable to sufficiently uniformly plasticize the material.

When a molding machine is purchased, then the buyer is offered a choice of screw sizes; three are usually available and these may be referred to as A, B or C. They have different screw diameters and allow an appropriate match of shot size (the feed system volume plus the molding volume) to barrel capacity. However, because they have different screw diameters, not only will the shot capacity vary but so will other things, for example, injection pressure and plasticizing capacity. Normally 'A' type screws have the lowest shot capacity, and the highest injection pressure, whereas C type screws have the highest shot capacity and lowest injection pressure.

The maximum shot capacity, or shot volume, is given by:-

$$Q = 0.25\, \Pi d^2\, S\, C.$$

Where, Q = the maximum shot capacity in cm^3 or, in^3;
Π = 3.142
d = the screw diameter in either cm or inches;
S = the screw stroke in cm or, inches;
C = the volumetric yield of the screw. If not known use 0.85%.

20 MELT TEMPERATURE

This is as measured in the nozzle or by an air shot technique. If measured by an air shot technique then, great care should be taken during the measurement so as to ensure that purging of the hot plastic does not cause an accident. This is because a plastic melt is hot enough to cause severe burns and it also adheres to the skin; burns are a common injury in injection molding shops. For this reason it is recommended that gloves and face shields are worn when handling hot material or where there is any danger of being splashed with plastic melt, for example, during start-up or purging. To obtain greater accuracy, when measuring the an air-shot, the tip of the measuring probe should be pre-heated to the temperature that it is expected to measure.

It is usual to find a temperature range quoted for each material with the upper end being used for injection moldings which have thin sections: the lower end of the range is used for heavier sections or, when the molder is inexperienced in the use of that particular material.

For each material a target melt temperature will be given. For a given machine, the actual injection cylinder settings to achieve this melt temperature, will depend on, for example, the screw rotational speed, the back pressure, the shot size and the molding cycle.

21 BARREL RESIDENCE TIME

The rate of decomposition of plastics, is dependent on both temperature and time. For example, a plastic may be degraded by a short exposure to a high temperature or, by a longer exposure to a lower temperature. How long the plastic is in the injection cylinder is therefore important. With some materials, for example ABS, the injection moldings may appear satisfactory on ejection but, on storage, colored areas (yellow/brown) may develop if the residence time is too long.

The actual residence time may be determined experimentally by measuring the time taken for colored plastic to pass through the injection cylinder. May be calculated very roughly by the following formula;

$$\frac{\text{rated capacity of injection cylinder (g or oz)} \times \text{cycle time (s)}}{\text{shot weight (g or oz)} \times 30.}$$

As a general practice, the residence time for a material should be calculated, particularly when using small shot weights on large machines, as the material can suffer degradation which is not always visually detectable. Because of the importance of residence time an approximate idea of it at a specified temperature will be given where possible.

22 TEMPERATURE SETTINGS

It should be noted that it is the melt temperature which is important and that the injection cylinder temperatures quoted are only guidelines. If you have no experience of processing a particular grade of material, then start with the lowest settings. (For ease of control, injection cylinders are divided into zones). All temperatures may be set to the same value; if the cycle times are long or, if working at high temperatures, set the first zone temperature to a lower value; this will prevent premature melting and bridging.

To ensure that the material freely enters the feed section of the screw, on most machines, cooling water is passed around the hopper throat assembly. This removes the heat from the throat but, it also removes a great deal of heat from the first, heated zone. Cooling water is also used to extract surplus heat from the hydraulic oil. (Injection molding should only begin when the temperature of the hydraulic oil has reached its operating temperature, for example, 45°C or 113°F. On many machines the oil is preheated to this temperature; this gives more consistent moldings).

The flow rate and temperature of the cooling water (to the throat and, to the oil cooler) are important as these variables affect how the material is plasticized within the barrel and, how the hydraulic components function. Usually the water used is from the mains supply and is therefore subject to a seasonal temperature variation from winter to summer. This variation, and the variation in flow caused through using an unmonitored supply, can cause product variation.

In order to overcome this problem, the use of an indirect, temperature controlled, type cooling system (such as a mold temperature controller) should be considered with the flow rate, and the temperature of coolant, standardized for each machine. Alternatively fit a solenoid actuated valve into the coolant supply: this should only permit water flow when the temperature exceeds a preset value.

The temperature of the material being fed to the injection molding machine should also be reasonably consistent.

23 INJECTION SPEED

This refers to the speed of mold filling, that is when the screw is acting as a ram. When molding thin sectioned components, high injection speeds (sometimes called "screw displacement rates" or, "plunger injection speed" or, "ram speed") are essential in order to fill the molding before freezing occurs. However, a better surface finish is obtained on moldings with thicker sections by using a slower speed. The initial speed of mold filling is usually complete, on most injection molding jobs within 1second. It must not be forgotten that the same ram speed on two differently sized machines, will produce different melt front velocities in the feed system and within the mold. It is the local melt front velocity that directly affects properties. Increasing the number of gates will also decrease the local melt front velocity. Fast filling tends to put more orientation on the part surface and less in the core of the molding. Conversely, slow filling minimizes surface orientation and can increase core orientation – usually gives more uniform orientation throughout the component.

Many molding faults, for example jetting and air trapping, may be avoided by using a range of speeds (that is, programming the injection speed), during the mold filling stage. May be done under open loop or closed loop control.

Many modern injection molding machines have closed loop control of the injection molding stroke. A sensor is used to track the screw position (often referred to as ram position) and, the position of the screw with respect to time can be plotted. The information from the sensor is also fed to a controller and this asks for progressively higher injection (line) pressure so as to maintain a constant injection molding rate. The pressure progressively increases because the resistance to flow progressively increases as the mold fills. At a certain point in the injection molding stroke, for example, when the mold is almost full or when the gate freezes, the resistance to flow becomes very high and, it becomes unrealistic to expect the screw to maintain the desired rate. At this point control is shifted from being velocity controlled to being pressure controlled: This point is known as Velocity Pressure Transfer or, VPT.

N.B. No matter what speed is used it should be recorded on the setting record along with the injection time. This is the time taken to reach the pre-set, first stage, injection pressure and is part of the screw forward time (SFT).

24 INJECTION PRESSURE

During the mold filling part of the molding cycle, high injection pressures may be needed in order to maintain the desired mold filling speed. Once the mold is full then this high pressure may not be necessary, or even desirable. In many cases, a high first stage pressure may therefore be followed by a lower, second stage pressure. When molding some semi-crystalline, thermoplastics materials, for example nylon and acetal. then the use of second stage pressure may not be required as abrupt changes in pressure can cause undesirable changes in crystalline structure.

To decrease the level of orientation in the molding it is important that the mold is filled as quickly as possible and that the plastics melt is not sheared while it is being cooled- what is called cold, creeping flow. Such an undesirable situation could result if the VPT point was set at the wrong position, for example, if control was handed over, from velocity to pressure, too soon. If the gate was still open, and if the packing pressure was only high enough to give slow filling then, a high level of internal stress in the molding would result: this is because the level of retained orientation is being maximized by cooling the molding while filling slowly. In such a case it would be better to increase the packing pressure so as to raise the mold filling rate.

It is therefore very important that the VPT point is capable of being set very precisely and that it is also very reproducible. If these conditions cannot be met then molding with varying properties will result. As the object of injection molding is to produce moldings with identical properties any variation of component properties is obviously unwelcome.

The VPT is capable of being set by the machine operator provided that he/she is given appropriate monitoring equipment. Changeover at the VPT may be set or triggered in the following ways:-

i) Screw position- also known as path dependent switchover;
ii) Hydraulic pressure- also known as line pressure;
iii) Nozzle pressure- also known as melt pressure;
iv) Cavity pressure- also known as CPC;
v) Mold opening force;
vi) Mold opening position.

Of the various VPT options currently available, CPC appears to be the most suitable as control is exercised from where it is required, that is, from within the mold: the pressure in the mold controls, for example, molding shrinkage and component weight. However, CPC has not made the impact expected because of reasons such as, the mold modification required (many trade molders have a large number of molds), the incorporation of the transducer within the ejection plate can be difficult, setting difficulties, transducer calibration problems (for example, caused through there being a variation in ejection pin size), ease of transducer damage and, the high cost of repair. On fast cycling injection molding machines, the machine hydraulics cannot respond quickly enough and so the VPT transfer is not effective when CPC is used.

Many molders now believe that there has been a significant improvement in machine design and operation in recent years; for example, they believe that machines are easier to set and are more reproducible. Because of this change it is believed by most that there is no longer a need for such sophistication as CPC control except, perhaps, for technical moldings. Some machine purchasers insist on having the CPC option just in case it is required at a future date; in many cases it is never used because it is thought there is no need to use it or, because there is no one who can set it up properly.

If control must be exercised from anywhere, apart from screw penetration, then most molders will go for line pressure control as this is easier, simpler and the system is less prone to damage; it is also virtually maintenance free. The use of mold parting, which occurs on injection, is now a commercial option. This is a very easy system to fit as, no machining is required as the sensor location is on the outside of the mold. However, it is probable that with the development of suitable nozzle pressure transducers that, control will be exercised in future from this part of the machine.

25 SCREW ROTATIONAL SPEED

Some of the heat necessary to plasticize the plastics material comes as a result of rotating the screw; the faster it is rotated, the higher is the temperature. Although high screw speeds are possible, it does not follow that a high speed should be used. It is better to adjust the rotational speed to suit the injection molding cycle by, for example, reducing the speed of rotation to the lowest value possible. This will give more uniform temperatures, reduce the wear and tear on the machine and, reduce the residence time at the front of the injection cylinder. Because of the importance of screw surface speed, the screws on larger machines should be rotated more slowly than those on smaller machines: this will keep the shear rate down and stop localized overheating. The speed of rotation will also differ from material to material. Rigid PVC, UPVC, is limited to a maximum of $0.25 ms^{-1}$ or 0.25m/sec, (15m/min or, 49ft/min), whereas with other more shear resistant materials, higher speeds can be employed.

To convert from rpm to ms^{-1} (m/s), multiply the rpm by the screw diameter in mm and then by 0.0000524. For example: to determine the actual screw speed for a 30mm diameter screw, so as to achieve a screw surface speed of $0.25 ms^{-1}$ or 0.25m/sec, the procedure is as follows:

$$\text{screw surface speed} = \text{Diameter of screw} \times \text{screw speed} \times 0.0000524$$
$$\text{where } 0.25 = 30 \times \text{screw speed} \times 0.0000524$$
$$\text{hence the screw speed} = 0.25 / (30 \times 0.0000524)$$
$$= 159 \text{ rpm}$$

26 BACK PRESSURE

This is the pressure that the screw must generate, and exceed, before it can move back. The use of high back pressures can improve color dispersion and material melting, but it is paid for as it increases the screw retraction time, reduces fiber lengths in filled systems and, imposes stress on the injection molding machine. Keep as low as possible and in any event do not exceed 20% of the machines (maximum rated), injection molding pressure. Adjust the back pressure, and the screw rotational speed, so that the screw comes forward again as soon as possible after the plasticization stage.

When structural foams (SF) are being injection molded, then the back pressure must be higher than the gas pressure which is being developed otherwise the screw will be pushed out of the barrel without doing any work.

On some machines it is possible to program the back pressure so as to compensate, for example, for the effective reduction in screw length which occurs during plasticization; such a reduction means less heat input and therefore a drop in temperature. However, on many machines it is difficult to sensibly set the machine as there is no easy way of measuring the effects of the changes.

27 SHUTTING DOWN

During a temporary stoppage, periodically purge the cylinder, or barrel, by passing material through the machine and/or by making air shots. If the plastics material looks discolored then increase the frequency of this purging. During a minor repair, the heaters on the injection cylinder should be set to low values, for example 150°C or 302°F, so as to minimize thermal degradation.

If a thermally stable plastic (for example, PS or PE) is being injection molded, then for an overnight stop, it is only necessary to close the slide at the base of the feed hopper, turn off the cylinder heaters (leave the nozzle heater on) and then purge the cylinder clean by pumping the screw dry. When nothing more comes from the nozzle, put any barrel cooling on maximum and when the machine is cool, turn everything off. The machine is then ready for reheating.

When high cylinder or barrel temperatures are used, then modify the above procedure slightly so as to prevent thermal decomposition of the resin. For example, turn off the cylinder heaters (leave the nozzle heater on), put any barrel cooling on maximum and then, periodically pump resin through the machine while it is cooling. Close the slide at the base of the feed hopper and purge the cylinder clean by pumping the screw dry and/or, by making air shots. When nothing more comes from the nozzle, and when the machine is cool, turn everything off. The machine is then ready for reheating.

Decomposition, or burning, of the plastics material in the injection molding machine will cause color changes and will then result in the subsequent injection moldings being rejected. When this happens a complete shut down and clean out may be necessary. To prevent this it may be necessary to purge a heat sensitive resin with another, more heat stable, plastics material as this will withstand subsequent reheating.

Some plastics degrade relatively easily, for example POM and PVC. If there has been an emergency shut-down while processing such materials then, before further injection molding is attempted, put the nozzle heater on *and only when the plastics material is melted in the nozzle, switch on the cylinder heaters.* Set the cylinder, or barrel temperatures to approximately 140°C, or

284°F, and allow the cylinder to equilibrate at this temperature. Then raise the temperature to the processing temperature and purge clean as soon as possible. All purgings should be put in a bucket of cold water. If a change is being made, from or to, another plastic, then purge with a natural, non-flame retardant grade of PS or PE. NEVER mix POM (acetal) and PVC (vinyl) or, follow one with the other without thorough purging with PS or PE.

Always check that you have followed the correct procedure before you turn a machine off or, before you change to another material. Material suppliers issue leaflets which give a great deal of information and are readily available. Such leaflets should be studied and an 'official shut down procedure' devised from them for each material. This should then be placed by the machine so that it is readily available to the machine operator.

28 REPROCESSING

The ability to reclaim faulty moldings, and the feed system, produced during the molding of thermoplastic materials, is very often a mixed blessing as it can lead to the attitude that the production of faulty moldings does not matter. 'After all, the molding can be ground up to give reclaimed material and so used again' is an often heard comment.

Of course they can be reclaimed but think of the wasted time and effort: it costs an awful lot of money to keep an injection molding machine running, so it makes sense to try and get molding production right first time.

When faulty moldings, and feed systems, are scheduled for reclamation then they should be looked after very carefully and, only those parts which are free from contamination and color changes should be used. This is because it is common practice to blend virgin (new) material with reclaimed material (regrind) and if the reclaimed material is contaminated, then a lot more reject material is produced. If unchecked the problem can quickly get out of hand. So, ruthlessly reject any molding, or any feed system, which is suspected of being contaminated, i.e. only reclaim good quality material. Scrap any material purged from the injection cylinder.

Keep all material, which is to be reclaimed, covered. Keep the grinder, and the storage containers, spotlessly clean. Inspect the grinder blades regularly for breakage, bluntness, and wear and then, replace when necessary. Put the reclaimed material through equipment which will remove dust or fines, metal fragments and water. Then treat this material as new stock, for example, store the reclaimed material in tightly, sealed containers in a clean, dry storeroom. Blend with virgin material in a precisely agreed ratio and, ensure that this ratio is adhered to during production as otherwise inconsistent molding will result.

Regranulation of feed systems can be carried out at the molding press by means of an auger regranulator assembly. The entrance section of the regranulator is positioned adjacent to the chute of the molding machine into which, the feed system falls when ejected from the mold. Once the material has been regranulated, it is then automatically blended at the required ratio with virgin material and fed directly back into the molding machine hopper. This type of process has two distinct advantages over a separate system; it minimizes the risk of contamination and the need to redry hygroscopic materials: however, this type of assembly does take up invaluable floor space around the molding machine.

On some jobs the use of reclaim cannot be tolerated and so particular care should be taken to minimize its generation.

29 FINISHING

Many plastics moldings are produced which require little or no finishing; with many other molding jobs, the only finishing operation is one of removing the feed system. However, many components are decorated and where appropriate, comments on relevant finishing processes will be included in this section.

30 OTHER COMMENTS

In this section any relevant comments, not covered in other sections, on the molding of a particular plastics material will be included. When considering the molding details of an individual material it is all to easy to forget the object of injection molding. The object of injection molding is to produce injection moldings of the required quality, in the right quantity, at the right price and, at the right time. This can only be done by careful consideration of each part of the production process. This careful consideration is becoming even more crucial with the growing use of fully automated, injection molding as robots, pick and place devices, conveyors, assembly machines etc., depend on a constant flow of injection molded components.

The importance of careful and accurate records cannot be over-emphasized.

31 TYPICAL COMPONENTS

Thermoplastics materials are used because injection molding to tight tolerances of consolidated components is possible (consolidated means that several non-plastic components have been replaced by one injection molding). This gives economies and design freedom. Such features are enhanced by the ease of joining moldings using both recoverable and non-recoverable joints.

The strength, stiffness, heat resistance and, corrosion resistance of many plastics must also not be forgotten. All these benefits are lost unless the components are sensibly designed. Computer aided selection of materials, followed by computer aided design of the components, can be a tremendous help in this respect.

To help illustrate the use and application of a particular material, a range of moldings produced from the plastic under discussion (together with some reasons for its selection) will be mentioned in this section. The list will not be exhaustive but will highlight typical components

32 UNIT PRICE

The price given will be based on the assumption that the material is purchased in 1 ton/tonne lots; a tonne is 1,000 kg. If purchased in this way then the plastics material will probably be supplied in 25 kg/55.1lb sacks, with 40 sacks shrink wrapped onto a pallet. When supplied in 50lb/22.6kg sacks, then the material is often supplied in 20 sack (1,000lb) loads. Supplying the material in sacks, or bags, is not the only way that the material can be supplied: considerable discounts may be obtained if the plastics material is purchased in bulk and, supplied in, for example, bulk containers. With some materials, this form of supply may not be acceptable as, for example, excessive water absorption may occur.

The price for each material will be given as a comparison to the price of a general purpose material; where the price of the general purpose material = 1.0. As plastics materials are bought by weight and sold by volume (in the form of moldings) then, any material with a low density, or specific gravity (SG), has a tremendous advantage. Often therefore, the cost per unit volume is more important than the cost per unit weight. To convert to the cost per cubic inch (cu in), multiply the cost per pound (lb) in cents by the SG and then, by 0.0361.

ETHYLENE VINYL ACETATE COPOLYMERS

1 Common name

Ethylene vinyl acetate.

2 Abbreviation(s)

EVA ; VAE; EVAc

3 Alternative name

Ethylene vinyl acetate copolymer(s).

4 Some suppliers   5 Trade names or trade marks

Atochem	Evatane
BASF	Lupolen KX
Bayer	Levapren
Dupont	Elvax
Enimont	Riblene
Exxon	Escorene/Escorene Ultra
Neste	Neste EBA/EVA
Petroquimica Triunfo	Trithera

6 Material Properties

Despite the fact that copolymerization with vinyl esters, such as vinyl acetate, was studied in the very early days of polyethylene polymerization, it is only comparatively recently that these materials seem to have made an impact in injection molding. Polymerizing vinyl acetate (VA) with ethylene, can disrupt the crystal structures that are present in polyethylene (PE) and will eventually give an amorphous, thermoplastic material. By varying the percentage of VA in the composition, polymers with significantly different properties are produced. As the percentage of VA is increased the transparency and flexibility of the copolymers increases. Injection molding grades of EVA contain from 4 to 30% of VA and such materials are tough, semi-opaque thermoplastics with a comparatively low, upper working temperature, for example, 65°C (149°F). EVA is softer, clearer and more permeable than LDPE; has better environmental stress cracking resistance, is tougher and will accept fillers more readily. Some grades of EVA can retain their flexibility at temperatures as low as $-70°C$ ($-94°F$).

Can be crosslinked by peroxides and/or, may also be chemically blown to give cellular products; the density of crosslinked, foamed EVA can be very low indeed and is far lower than that possible without crosslinking. Such crosslinked, foamed EVA can have a very fine structure similar to micro-cellular rubber. The inherent flexibility, good processing characteristics, low odor and no plasticizer migration problems make this material an attractive alternative to PPVC, natural and synthetic rubbers.

7 Ease of Flow

Available in a wide range of grades but can be a very easy flow material; MFI (MFR) can range from 2 to 35. As usual, higher molecular weight grades have the worst flow.

8 Shrinkage

Depends on VA content – as this increases, crystallinity and shrinkage decreases. Is of the order of 0.5 to 4 in/in or mm/mm ie 0.5 to 4%. Shrinkage values of as low as 0.5 to 1.5% can be obtained from specific grades. In general, EVA's wide shrinkage range makes it extremely difficult to accurately predict the shrinkage value – necessary to achieve the desired dimensions for a particular component. When planning to use this material, a sample cavity is manufactured and moldings are produced using typical processing conditions (i.e. production temperatures, pressures and cycle time). The shrinkage values obtained from these moldings, would then be used to size the cavity and core dimensions of the production mold.

9 Resistant to the Following

Flex cracking at low temperatures and also to ozone cracking. Disinfectants, ethylene oxide and sodium hypochlorite may be used for cleaning or disinfection. Chemical resistance is similar in many respects to LDPE but not as chemically resistant. Low MFR grades have the best resistance.

10 Not Resistant to

Steam sterilization – use gamma irradiation if sterilization is required. Resistance to aromatic and chlorinated solvents is poor. Soluble in aromatic and chlorinated hydrocarbons at elevated temperatures.

11 Material Detetection or Identification

With a density of 0.926 to $0.95 gcm^{-3}$ (0.537 to 0.551 oz./cu.in.) the material (solid, non-filled material) will float in both water and in saturated, magnesium chloride. The natural color of the material is similar to that of LDPE at low VA content (i.e.

an off-white material) but at higher VA contents (eg 20%) the material is colorless.

The natural color of the material ranges from milky white to colorless. When this material is heated in a flame it ignites easily and burns with a yellow-tipped, blue flame giving off only a little smoke; forms burning drops and when the flame is extinguished, gives a smell like candle grease (high vinyl acetate content EVAs can also emit a smell of acetic acid, however this acidic aroma is often difficult to detect). When heated in the absence of a flame the material will soften and melt to give a clear liquid, as the crystal structures are destroyed. Is stable in the absence of air up to approximately 240°C (464°F) when it decomposes to give low molecular weight hydrocarbons. Can be cut easily with a knife and can also be easily scratched with a knife or fingernail. This material dissolves with difficulty in solvents such as xylene, carbon tetrachloride and trichloroethylene. This material will swell in solvents such as ethyl acetate, petrol and methyl ethyl ketone (MEK).

12 Coloring

As the natural color of the material ranges from off-white to colorless, then a wide color range is possible; this includes both transparent and opaque colors. Sold in both compounded colors and as natural material for coloring on the injection molding machine by techniques such as dry coloring, masterbatching and liquid coloring. When coloring EVA, organic dyes should not be used due to the problems of color leeching or bleeding. For most coloring purposes, the use of organic and inorganic pigments are preferred. When dry coloring EVA, colorant levels of up to, and including 1%, are generally used. Wetting agents are used with dry colorants primarily for cleanliness in the molding shop rather than aiding the dispersion of the colorant. Universal type masterbatches are often used with EVA at concentration levels of 1%: however to obtain a more uniform color on the molded component the addition level may need to be increased to 3 to 5%. When color uniformity is particularly important, EVA based masterbatches are often preferred. If opacity of color is required, then inorganic pigments tend to be used. There are no real limitations to using liquid colorants with EVA; however, colorant addition levels have to be kept to a minimum (i.e. no more than 3%) otherwise screw slip can occur.

More often than not, molders use the same approach as for coloring GPPS; however, the significant difference between PS and EVA is the softness of EVA and its very easy flow characteristics. To improve the dispersion of the pigments within the melt, it is necessary to create a adequate mixing/grinding action within the barrel of the machine. In order to achieve this, it is often necessary to reduce the temperatures on the machine to the lowest value that will still enable components to be produced of the required quality.

13 Materials and Components Handling

Will absorb less than 0.2% water in 24 hours at room temperature. This means that drying is not normally necessary. If it is, dry in a hot air oven for 8 hours at 50 to 60°C (122 to 140°F), in a desiccant dryer for 4 hours at 50 to 60°C (122 to 140°F). The actual drying temperature is dependent upon the amount of vinyl acetate incorporated, for instance, a drying temperature of 50°C (122°F) should be used for materials that contain 33% or more vinyl acetate and 60°C (140°F) for compounds that contain less than 33%.

Normal material handling techniques such as a hopper loader are readily used with EVA. However, when pneumatically transporting this material over long distances the design of the handling system must accommodate the low surface hardness and the low softening point of EVA.

14 Mold and Gate Considerations

A suggested mold temperature is 25°C (77°F). If a melt temperature of 200°C (392°F) is used then the maximum amount of heat that needs to be removed for each gram of shot is 563 Joules(J). Over the temperature range quoted, the specific heat is approximately 3130 J/kgK.

Typical runner sizes range from 4mm to 7mm (0.157 to 0.276 in.) in diameter. Diameters above 7mm (0.276in) tend only to increase the cycle time and do not significantly assist in the filling of the mold. Runner lengths should be kept as short as possible and suitably sized ejectors positioned at runner junctions to ensure that the feed system is removed from the mold without difficulty.

All types of gates are used for EVA. However, one of the main criteria for gate selection, is the ability to be removed from the molded component without leaving unsightly gate scars or any vestige: the gate must not introduce excessive melt turbulence during mold filling. As a result of the ductile, or elastic nature, of EVA, most types of gates are usually removed from the component by a separate degating operation. Typical degating methods used are cropping, cutting by means of scissors or scalpel, or by simple twisting. Fan, film and tab gates are preferred to other types so as to reduce jetting and other surface blemishes. Gate lands for the edge, pin or film type gates should not be greater than 0.75mm (0.030in) in length. Submarine type gates and reverse tapered sprue pullers are widely used for cold runner molds, intended to run fully automatically. To ensure that minimal gate vestige occurs, when using a submarine gate, an inclusive gate angle of 30 degrees and a correctly positioned ejector pin (i.e. not greater than 10mm (0.394in) from gate entry) are essential.

EVA is ideally suited for runnerless type molds – due to its ease of processing and relatively good thermal stability. All types of runnerless systems can be used for EVA; however, for production runs that require frequent color changes, the insulated hot-tip type, runnerless system is preferred. Typical gate diameters used for runnerless molds are 1 to 2mm (0.039 to 0.080in). Accurate temperature control of the runner manifold and nozzle assembly is essential, particularly when using melt temperatures of 220°C to 230°C (425°F to 446°F); poor temperature control can often lead to the emission of fumes, including acetic acid, resulting from the degradation of the polymer. The position of the cartridge heaters in relation to the runner channel is important, otherwise degradation can occur as a result of the presence of a hot spot. When selecting the design of a hot runner manifold it is advisable to also seek the advice of the material supplier as well as the hot runner specialist.

If cold runner molds are used in preference to runnerless molds, for economic reasons, then a hot sprue bushing should be incorporated in the mold; so as to reduce the material content within the feed system and allow faster cycle times to be achieved.

Particular attention should be given to the design of the cooling circuitry in the mold when molding EVA. To keep cycle times to a minimum, and to prevent components from sticking in, the mold, surface temperatures should be controlled to 35°C (95°F) or below. However, mold surface temperatures of lower than 8°C (46°F) can cause excessive frozen-in stress to occur in the molding. To minimize component deformation, it is preferable to keep both mold halves at the same surface temperature. The position of the water channel in relation to the mold surface is very important when designing molds to be used with EVA. Because of the large amounts of heat that need to be removed from the mold, in order to achieve economic cycle times, the diameter of the water cooling channel should not be less than 8mm (0.315 in) and the distance from the mold surface to the edge of the cooling channel should not be greater than 1.5 times the cooling channel diameter.

Regranulation of the feed system of a cold runner type mold may be accomplished by using a work handling device, or sprue grabber, so that the feed system is fed directly into the regranulator. Use of this technique ensures that minimal contamination occurs during the regranulation operation.

The rubbery nature of EVA should be considered when designing the ejection system. Ejection of thin walled and cylindrical components should be carried out using stripper plates, rings or a sleeve so as to maximize the available surface area on the cavity or core. Wherever possible, apply a matt surface to the core by vapor, or shot blasting so as to enable ease of ejection. Draft angles of between 2 to 5 degrees per side are recommended. The higher the EVA content in the polymer, the greater the draft angle required to eject the component. If difficulties arise with ejection of the component it is advisable to use a lubricated grade of EVA. The lubricant (0.5% butyl stearate) or slip/antiblock additive (oleamide or stearamide) can either be added in the form of a powder or masterbatch, for example, by tumble blending. Other release agents, e.g. zinc stearate or PTFE spray, can be applied to the surface of the mold so as to prevent component sticking. However, the use of silicone based release agents should be avoided as stress cracking can occur to the moldings.

Mold venting is essential when using EVA, as the filling pattern of the mold can be considerably altered with poor or non-existent venting; typical sizes are 0.025 to 0.038mm (0.001 to 0.0013in) deep by 3mm (0.118in) wide. As EVA tends to flow very easily (more easily than PE-LD) it is essential that the correct size vents are used otherwise flashing will readily occur.

Threaded components can be readily stripped off, or undercuts bumped off, as a result of the elastic nature of EVA; wherever possible always round off the edge of threads or undercuts so as to ease extraction and prevent damage occurring to the component: typical depths of undercuts are 0.2 to 0.5mm (0.008 to 0.020in).

Transparent and opaque moldings that require a glossy surface necessitate the mold surface to be of a mirror finish. However the higher the gloss, the more difficult it is to eject the molding (due to sticking) and for this reason a matt or textured surface finish is preferred. Matt or textured finishes also disguise the flow marks that may occur during molding and/or the scratches incurred while in service.

Conventional steels of the nickel-chromium alloyed type are used for production molds where surface hardnesses of up to 52 Rockwell C are necessary. To prevent corrosion, and to extend the tool life, for long production run molds, it is preferable to use a high chromium content steel (i.e. 13% or greater).

15 Flow Path: Wall Thickness Ratio

For general purpose grades this is approximately 200 to 250:1 but it can reach 290 to 300:1 for easy flowing grades.

16 Projected Area

Usually 1 to 2.5tsi. (15 to 38.5MNm^{-2}) is sufficient.

17 Cylinder Equipment

General purpose screws can be used, for example, with an L:D ratio of about 20:1 and a constant pitch of 1D. The feed zone should be approximately 0.5L, the compression zone 0.25L and the metering zone 0.25L; a compression ratio of between 2.5 to 3.5:1 will be sufficient. The screw is usually fitted with a non-return valve but, valves are not essential on the nozzle unless high back pressures are involved. The best results have been obtained with spring loaded or hydraulically operated types; decompression or suck back is also used.

Shallow flight depth in the metering section of the screw may cause excessive shear thus making melt temperature control difficult. Typical flight depths for the feed and metering sections are as follows:

Screw Diameter	Feed section	Metering section
40mm (1.57in)	7.6mm (0.299in)	2.2mm (0.087in)
50mm (1.97in)	8.5mm (0.355in)	2.5mm (0.098in)
65mm (2.56in)	8.8mm (0.346in)	2.7mm (0.106in)
75mm (2.95in)	9.5mm (0.374in)	2.9mm (0.114in)
90mm (3.54in)	10.0mm (0.394in)	3.3mm (0.130in)

18 Screw Cushion

On smaller machines this is approximately 3mm (0.118in); on larger machines 9mm (0.354in) is suggested.

19 Shot Capacity

Because of the material's relatively good heat stability as little as 20% of the cylinder's rated capacity may be used: the maximum possible is 90%. It should be noted that the melt density of this material is 0.7 to 0.74g/cm^3 (0.42 to 0.44cu.in) compared to 0.95g/cm^3 (0.551oz/cu.in) for PS; therefore the quoted shot capacity for a particular screw and barrel assembly needs to be downrated by 24%. For example, if the shot capacity in PS is 220g, then the actual shot capacity when molding EVA would be 169g.

20 Melt Temperature

This is as measured in the nozzle or by an air shot technique; 140°C to 225°C (284°F to 437°F) is usually suggested. the actual melt temperature is dependent upon the grade of EVA being molded. However, if melt temperatures are too low there is a danger of high residual stresses and subsequent environmental stress cracking.

21 Barrel Residence Time

At a temperature of 225°C (437°F) the maximum residence time in the barrel should not exceed 5 to 6 minutes. When the temperature is 235°C (455°F) then the maximum residence time in the barrel should not exceed 2 to 3 minutes.

22 Temperature Settings

Low VA content, for example, 7.0%

Zone No.	Location	Temperature in °C		Temperature in °F	
		FROM	TO	FROM	TO
0	Hopper throat	30	50	86	122
1	Barrel rear (hopper end)	130	160	266	320
2	Barrel middle	150	180	302	356
3	Barrel middle	170	190	338	374
4	Barrel front	180	210	356	410
5	Nozzle	190	220	374	428
6	Mold	15	40	59	104

High VA content, for example, 28%

Zone No.	Location	Temperature in °C		Temperature in °F	
		FROM	TO	FROM	TO
0	Hopper throat	30	50	86	122
1	Barrel rear (hopper end)	120	140	248	284
2	Barrel middle	130	150	266	302
3	Barrel middle	140	160	284	320
4	Barrel front	150	170	302	338
5	Nozzle	160	180	320	356
6	Mold	15	30	59	86

Please note that it is the melt temperature which is important; those temperatures given in the table are only suggested, initial settings. The temperature of the hydraulic oil, the hopper block and the material temperature should also not vary excessively. Mold temperature control is extremely important and therefore these temperatures must be checked regularly.

23 Injection Speed

Moldings requiring a high surface gloss and good mechanical properties are molded using slow injection speeds. Fast injection speeds causes melt turbulence to occur in the region of the gate producing a rough surface finish on the molding. The extent of melt turbulence is related to the type and size of gate used. Profiled injection speeds are also used to overcome such problems, particularly when molding large surface area components.

24 Injection Pressure

The machine should be capable of giving up to:

first stage, up to 1500bar; 150MNm^{-2}; 21,750psi.
second stage, up to 750bar; 75MNm^{-2}; 10,850psi.

Careful control over dwell time and pressure is needed in order to avoid overpacking but yet still be able to compensate for shrinkage.

25 Screw Rotational Speed

Adjust speed (the rpm) to suit the molding cycle but do not use a screw surface speed of greater than 0.5 to 0.55ms^{-1} (1.64 to 1.8ft/sec) If a screw of 40mm (1.57in) diameter is being used then the rpm should not exceed 262. Similarly the maximum diameter of screw when rotated at 200rpm, in order to obtain a screw surface speed of 0.5 to 0.55ms^{-1} (1.64 to 1.80ft/sec), is 52mm (2.05in).

26 Back Pressure

Up to 150bar; 15MNm^{-2}; 2,175psi. Should be kept as low as possible: usually a nominal value of 5 bar (0.5MNm^{-2}, 72.5psi) is used so as to achieve consistent shot weights. May need to be increased, to assist dispersion, when using colorant but keep as low as possible.

27 Shutting Down

No need to purge with another material. If melt temperatures close to the upper limit are being used, it is preferable to switch off the heaters and continue purging until the set temperatures can be seen to fall. Because of the low molding temperatures used for some grades of EVA, and its softness, it is preferable to use PE-LD or PE-HD to purge the cylinder clean.

28 Reprocessing

When this material is reclaimed, it is essential that the cutter blades are sharp and free from damage so that they cleanly cut this rubbery material. Care should also be taken, when using regrind, not to include material that has been previously degraded, for example, due to long barrel residence times. Typical regrind levels of up to 30% are used.

29 Finishing

This material can be readily decorated by hot foiling and silk screening; however, to achieve good results the use of lubricated grades should not be considered. Can be welded using the hot plate, radio frequency and ultrasonic techniques.

30 Other Comments

The weathering behavior of EVA is similar to that of LDPE and like LDPE, it can be significantly improved by the incorporation of 2 to 3% carbon black. Since the coefficient of friction is much higher than other polymers, EVA is used for applications where slippage needs to be overcome.

31 Typical Components

Initial uses for this material were as replacements for rubbers, LDPE and plasticized PVC. Flexibility does not depend upon plasticizers, which may be toxic, tainting or migratory, and so such materials are used to make record player, turntable mats. Also very useful in food packaging and surgical applications for the same reasons. It replaces rubber in some applications, because of its resistance to ozone attack, flexibility and low temperature and flex crack resistance, for example, in traffic signal hoods. Because of the low temperature flexibility of this material, it is used in place of LDPE for ice cube trays. Other applications include barrel bungs, bicycle saddles, railway sleeper pads, disposable baby bottle teats, ear tags for animals, refrigerator and freezer door gaskets, closure wads and WC pan connectors. Also used to make base pads for staplers and for small items of electrical equipment, When crosslinked, foamed EVA can be used to make cellular shoe soles and tires, for example, for push chairs and golf trolleys.

32 Unit Price

If the price of a natural, unmodified grade of GPPE is taken as one then the relative price of this molding material is

approximately, for a general purpose grade (5% VA content) 0.82: for a general purpose grade (28% VA content), it is 1.13.
The price relative to GPPS can also be found in Table 7.

HIGH DENSITY POLYETHYLENE.

1 Common Name

High density polyethylene: ultra high molecular weight, high density polyethylene.

2 Abbreviation(s)

HDPE; PE-HD; UHMW-HDPE.

3 Alternative Names

High density polythene, high density poly(methylene), low pressure polyethylene, linear high density polyethylene.

4 Some Suppliers   5 Trade Names or Trade Marks

Suppliers	Trade Names or Trade Marks
Alathon	DuPont
Allied	Paxon
Amoco	Amoco High Density Polyethylene
Atochem	Lacqtene HD
Bamberger	Bapolene
BASF	Lupolen HD
BP	Rigidex HDPE
Chemie Linz	Daplene
Dow	Dow HDPE
DSM	Stamylan HD
Enron/Delong Prochem	Norchem
Hoechst/Hoechst Celanese	Hostalen
Huls	Vestolen A
Enimont	Moplen RO
Neste	Neste
Novacor Chemicals	Novapol HD
Phillips	Marlex
Soltex Polymer Corp	Fortiflex
Statoil	Statoil HDPE
Union Carbide	HDPE.
USI (Quantum)	Petrothene
Wacker Chemie	Wacker Polyathylen

6 Material Properties

Under conditions of comparatively low temperature and pressure, ethylene can be polymerized to give a plastics material (HDPE) which is substantially free from branching. This is achieved by using stereospecific catalysts, for example, of the Ziegler-Natta type; these direct the incoming monomer and make the polymer chain grow in a very ordered way. Because of its regularity, and lack of chain branching, this material has a higher level of crystallinity than LDPE. This increase gives grades which have a higher density, rigidity, tensile strength, hardness, heat distortion temperature, chemical resistance, viscosity and resistance to permeability; however the impact strength is lower. Has good dynamic fatigue resistance but not as good as PP; some living hinge effect. As with all polyolefins, the weathering resistance is satisfactory but can be improved, for example, by the addition of carbon black. Compared to PP homopolymer, HDPE has better resistance to low temperature impact and to oxidation. PP has a higher Vicat softening point, better resistance to flexing, a higher hardness, a higher tensile and elongation and will also cycle faster: the gloss on moldings can be similar.

The water vapor permeability of this type of PE, as with all types of PE, is low. PE is permeable to gases and vapors (i.e. it lets them through) but LLDPE and HDPE are less permeable to gases and vapors than LDPE. Permeability for organic vapors is least for alcohols and then increases in the order shown; from acids to aldehydes and ketones, esters, ethers, hydrocarbons and halogenated hydrocarbons. (Permeability decreases with density). Some grades of HDPE are accepted as being suitable for containers for oil and petrol: they have been used for fuel tanks: in some cases the formed containers have been chemically modified, by fluorination or sulfonation, so as to make the material almost impermeable to fuels. HDPE will resist deformation by boiling water. The properties of HDPE are significantly controlled by it's MFR, density and molecular weight distribution values. Injection molding grades of HDPE are manufactured having a narrow molecular weight distribution compared to grades used for extrusion and blow molding.

When the density is 0.910 to $0.925g.cm^{-3}$ (0.53 to 0.54oz/cu.in), a PE material is sometimes known as Type 1; when the density is 0.926 to $0.94g.cm^{-3}$ (0.53 to 0.55oz/cu.in), it is sometimes referred to as Type 2. HDPE, with a density of 0.940 to $0.965g.cm^{-3}$ (0.55 to 0.56oz/cu.in), is sometimes known as Type 3.

The grades which have a density below 0.96/0.56 oz/cu.in are produced by using a second monomer at low levels (<1%). Strictly speaking they are therefore copolymers of PE with another α olefin, for example, with butene-1 or, with hexene-1. The use of the second monomer, reduces the density by introducing short, side chain branching. Such materials may be known as MDPE. HDPE/MDPE is widely used in the blow molding industry: approximately 55% of all plastics used in blow molding are HDPE/MDPE materials.

7 Ease of Flow

This material is an easy flowing material having MFR values ranging from 0.1 to 28. However, because the impact resistance of HDPE can be low when low molecular weight grades are used, then high molecular weight materials are preferred i.e. the materials have low values of melt flow rate (MFR). (The molecular weight of a HMW blow molding grade, as measured by solution viscosity, may be 150,000. That of an injection molding grade, or of a grade for small blow moldings, may be 75,000).

Impact strength falls off with increasing MFR (flow is becoming easier) and the plastic become more notch sensitive. HMW grades, in granular form, are now available for injection molding (Hoechst) and is said to be suitable for the injection molding of large and complex components.

MOLDFLOW DATA FOR BASF LUPOLEN HDPE

Material	Viscosity (Nsm^{-2}) at $1,000s^{-1}$ at the following temperatures			
	200°C (392°F)	220°C (428°F)	240°C (464°F)	260°C (500°F)
Easy flow grade (5431P)	82	72	63	55
Medium flow grade (6031M)	139	123	109	97
Stiff flow grade (5021D)	439	399	363	330

Material	Viscosity (Nsm^{-2}) at 240°C (428°F)			
	$100s^{-1}$	$1,000s^{-1}$	$10,000s^{-1}$	$100,000s^{-1}$
Lupolen 5431P	156	63	25	10
Lupolen 6031M	296	109	40	15
Lupolen 5021D	1584	363	83	19

8 Shrinkage

This is a semi-crystalline, thermoplastics material and so molding shrinkage is high; it is of the order of 0.015 to 0.04in/in or mm/mm ie 1.5 to 4%; may increase to 5% for thick sectioned moldings. Because of HDPE's wide shrinkage range it is extremely difficult to accurately predict the necessary shrinkage value in order to achieve the desired dimensions for a particular component. The shrinkage value is dependent upon the degree of orientation and crystallinity that occurs in the molded component. Because of this a sample cavity is often manufactured and moldings produced using typical processing conditions (i.e. those temperatures, pressures and cycle times used in production). The shrinkage values obtained from these moldings can then used for the sizing of the cavity, and core dimensions, of the production mold. However what must be taken into consideration is the degree of after-shrinkage that occurs with HDPE. What this means is that the sample moldings should be left for several days to condition before being measured. Mold temperature has a great effect on shrinkage and must therefore be controlled very precisely.

9 Resistant to the Following

No known solvent at room temperature; soluble in hydrocarbons and aromatic hydrocarbons at temperatures above 60°C (140°F). Such materials will also cause swelling at room temperature as this plastic is permeable to such materials. Also swollen by white spirit and carbon tetrachloride; in this respect HDPE is better than PP. Theoretically less resistant to oxidation than LDPE but in practice they are similar; could be due to catalyst residues.

10 Not Resistant To

PE is associated with environmental stress cracking; this occurs when a material is subject to strain, or an internal stress, in the presence of polar liquids or the vapors of such liquids. Such environmental stress cracking (ESC) is also associated with detergents or silicone fluids although there are many other environmental stress cracking agents, for example, chloroform, xylene and paraffin. Minimize the environmental stress cracking effect by reducing the residual stresses in the molding, by careful component design and by using the lowest MFR grade at any particular density level.
The light, or U.V., resistance is satisfactory; cheapest way of improving this is by black incorporation. Carbon black filled HDPE compounds are now being used for applications where very long term weatherability is required. Has limited resistance to oxygen at elevated temperatures; antioxidants are used for protection. Unless so protected the electrical properties will suffer. Resists aromatic and chlorinated hydrocarbons more than LDPE.

11 Material Detection or Identification

With a density of 0.94 to 0.965gcm^{-3} (0.55 to 0.56oz/cu.in) the material (solid, non-filled material) will float in both water and in saturated, magnesium chloride. The manufacturing processes used to produce PE are as follows.

Process	Product	Density (kg/m³)
Gas phase	LLDPE, HDPE	900 to 960
High pressure	LDPE	915 to 935
Modified high pressure	LLDPE, HDPE	880 to 940
Slurry	HDPE	935 to 960
Solution	LLDPE, HDPE	900 to 960.

(If a material has a density of 900kg/m^3, then the SG will be 0.9)

No solvent at room temperature but at higher temperatures (approximately 55°C or 131°F) soluble in hydrocarbons and chlorinated hydrocarbons, for example xylene and trichloroethylene. Below 60°C (140°F) approximately, PE is insoluble in all organic solvents but it does swell in aliphatic, aromatic and chlorinated hydrocarbons: the lower the density the more it swells.

The natural color of the material is a milky white and so a wide color range is possible; transparent moldings are not possible. When this material is heated in a flame then it ignites easily, burns with a yellow-tipped, blue flame and gives off only a little smoke; forms burning drops and when the flame is extinguished, gives off a smell like candle grease. LDPE has a melting point of 110 to 125°C (230 to 257°F), LLDPE 115 to 128°C (239 to 262°F), HDPE is 130 to 135°C (266 to 275°F) and PP is 165 to 175°C (329 to 347°F).

When heated in the absence of a flame then the material will soften and melt to give a clear liquid, as the crystal structures are destroyed, at approximately 135°C (275°F). Is stable in the absence of air up to approximately 300°C (572°F) when it decomposes to give low molecular weight hydrocarbons (cross-linked PE will not melt but will become rubbery at approximately 115°C (239°F)). Can be cut easily with a knife and can also be easily scratched with a knife or fingernail.

12 Coloring

As the natural color of the material is off-white then a wide color range is possible; this does not include transparent colors. Sold in both compounded colors and as natural material for coloring on the injection molding machine by techniques such as masterbatching, dry coloring and liquid coloring. When dry coloring, adhesion promoters such as paraffin can be used. Control the mixing time precisely as too short a mixing time gives poor dispersion and inadequate adhesion; too long a mixing time may cause pigment compaction. Do not exceed approximately 0.4% pigment concentration; higher color concentrations are achieved with colored compounds. Pigment selection is important as some types of pigments increases the sensitivity for stress cracking and/or warpage to occur to moldings. Consult the coloring specialists for advice on the type of pigments that can be used for HDPE. When using carbon black as a dry colorant care should be taken to prevent it from absorbing any moisture during storage otherwise the colored polymer will need to be dried before molding.

13 Materials and Components Handling

Will absorb less than 0.01% water in 24 hours at room temperature. This means that drying is not normally necessary. If it is, dry in a hot air oven for 3 hours at 65°C (149°F), in a desiccant dryer for 1 to 1½ hours at 80°C (176°F).

This material may not be joined to itself using solvents as there is no solvent at room temperature. Because of its inert, 'non-stick' surface it is difficult to bond using adhesives; some success with contact or hot melt adhesives. If the surface is made polar, for example by using a flame or an electrical discharge, then this material may be bonded to metals using epoxides or nitrile-phenolic adhesives; such treatments also improve printability.

14 Mold and Gate Considerations

A suggested mold temperature is 20°C (68°F). If a melt temperature of 240°C (464°F) is used then the maximum amount of heat that needs to be removed for each gram of shot is 800 Joules(J). Over the temperature range quoted, the specific heat is approximately 3640 J/kgK.

Runner diameters range from 4mm to 7.5mm (0.157 to 0.295in); a typical diameter is 6mm (0.236in). Diameters above 7.5mm (0.295in) tend only to increase the cycle time and do not significantly assist in the filling of the mold. Runner lengths should be kept as short as possible and suitably-sized, ejectors positioned at runner junctions to ensure that the feed system is removed from the mold without difficulty.

All types of gates are used for HDPE, however the main criteria for each type of gate is its ability to be removed from the molded component without leaving unsightly gate scars or any vestige. Most types of gates are usually removed from the component by a separate degating operation. Typical degating methods used are cropping, cutting by means of scissors or scalpel or, by simple twisting. Gate lands for the edge, pin or film type gates should not be greater than 0.75mm (0.030in) in length. Submarine type gates and reverse tapered sprue pullers are widely used for cold runner molds intended to run fully automatically. To ensure that minimal gate vestige occurs when using a submarine gate an inclusive gate angle of 30 degrees and a correctly positioned ejector pin (i.e. not greater than 10mm (0.394in) from gate entry) are essential.

HDPE is ideally suited for runnerless type molds due to its wide processing range and good thermal stability. All types of runnerless systems can be used for HDPE; however, for production runs, that require frequent color changes, the insulated hot-tip type runnerless system is preferred. Typical gate diameters used for runnerless molds are 0.75 to 0.8mm (0.030 to 0.032in), however gate diameters as low as 0.7mm (0.028in) are often used to prevent gate vestige occurring particularly when producing cosmetic components that require zero gate vestige. To prevent gate vestige, or a short fiber attached to the gate, it is best to use hot runner nozzles that incorporate a shut-off valve or needle.

Accurate temperature control of the runner manifold and nozzle assembly is necessary, particularly when using melt temperatures of $240°C$ ($464°F$), and above, so as to produce moldings of a consistent color. Poor temperature control can often lead to degradation of the pigment which results in color changes occurring during the production run. Gate temperature control is extremely important when molding HDPE as components have a tendency to warp or for cracks to appear in the gate region (due to a too low or too high a gate temperature). Too low a temperature can induce high levels of stress or orientation in the molding giving warpage, whereas too high a temperature causes overheating of the melt, or extended gate freeze times, which allows overpacking in the gate region to take place (causing cracks to occur on the molding while in service).

If cold runner molds are used in preference to runnerless molds, for economic reasons, then hot sprue bushings should be incorporated in the mold so as to reduce the material content within the feed system and to allow faster cycle times to be achieved.

Effective and accurate mold temperature control is of considerable significance when molding HDPE. What must be borne in mind when designing the cooling circuitry is the amount of heat that needs to be removed per gramme of shot. The figure of 800 Joules/gm (344 Btu/lb), already quoted, means that the mold will require twice the amount of cooling capacity (in order to remove the heat from the mold for HDPE), as for PS. Molds with poor cooling facilities require extended cycle times to prevent deformation of the molding upon ejection (i.e. the mold is being used as a cooling fixture rather than an efficient heat exchanger).

To prevent large shrinkage variations and/or warpage occurring to moldings of large surface areas, or in those of differing wall sections, the need for separate cooling circuits in the same half of the mold may be required. Each circuit should be independently controlled at a specific temperature, e.g. in the region of the gate a mold surface temperature of between 10 to 20°C (50 to 68°F) would be used, whereas a temperature of between 40°C to 60°C (104 to 140°F) would be used at the end of the melt flow path. By using different mold surface temperatures the overall cooling rate of the molding is much more uniform giving better dimensional stability. To obtain better cooling, particularly when molding thick sectioned components, certain areas of the mold should be made from beryllium copper in order to dissipate the heat more rapidly.

The position of the water channel in relation to the mold surface is very important when designing molds to be used for HDPE. In order to achieve economic cycle times the diameter of the water cooling channel should not be less than 8mm (0.315in) and the distance from the mold surface to the edge of the cooling channel should not be greater than 1.3 times the cooling channel diameter (1.3D). For moldings that possess both thick and thin sections, the distance of the water channel should be closer for the thicker sections (i.e. 0.9 to 1.0D) than for the thinner sections.

Molds should be suitably vented to accommodate fast mold filling speeds; typical sizes 0.025 to 0.08mm (0.0009 to 0.003in) deep by 6 to 10mm (0.236 to 0.394in) wide (where possible).

When molding HDPE, use the largest sized ejectors and maximum draft possible as HDPE has a tendency to stick to the surface of the mold. Typical draft angles of 1 to 2 degrees per side are recommended. Use the maximum size ejector pins wherever possible or, if inconvenient consider using stripper sleeves. rings or plates so as to obtain the maximum amount of surface contact area to eject the component.

To avoid excessive molded-in stresses resulting from the use of metal inserts always preheat the inserts to 80 to 120°C (176 to 248°F) prior to placing into the mold. All metallic inserts should be thoroughly degreased and have some form of undercut feature on their external surface (for instance a knurl or groove), but care should be taken not to create notch sensitivity.

When molding UHMW HDPE, the cross sectional area of the runners should be two to four times bigger than those used for conventional, injection molding grades of HDPE.

15 Flow Path: Wall Thickness Ratio

This material is an easy flowing material; at 1mm (0.039in) wall thickness the flow path: wall thickness ratio can reach up to 270:1 for a high MFR grade.(typical grade 200:1)

16 Projected Area

Usually 1.5 to 2.5tsi. (23 to 39MNm^{-2}) is sufficient. For components of thin wall sections and requiring a long flow length this may have to be increased to 3.5tsi or 54MNm^{-2}.

17 Cylinder Equipment

General purpose screws can be used ,for example, with an L:D ratio of about 20:1 and a constant pitch of 1D. The feed zone should be approximately 0.5L, the compression zone 0.3L and the metering zone 0.2L; a compression ratio of about 2:1 should be sufficient. The screw is usually fitted with a non-return valve as its use is essential when long screw forward times are needed so as to obtain good dimensional stability on the molding. Valves are not essential on the nozzle unless high back pressures are involved (may be necessary to ensure that good pigment dispersion occurs when using an on-line coloring technique). The best results have been obtained by spring loaded or hydraulically operated types; decompression, or suck-back, is also used.

The use of a short nosed torpedo design, for the back flow valve assy, instead of the long type is often preferred due to the high melt compressibility of this material. Such a design often reduces melt cushion and product weight variability.

18 Screw Cushion

On smaller machines this is approximately 3mm (0.118in); on larger machines 9mm (0.354in) is suggested.

19 Shot Capacity

Because of the material's good heat stability as little as 10% of the cylinders rated capacity may be used: the maximum possible is 95%. Due to the melt density of HDPE being 0.75g/cm^3 (0.435oz/cu.in.) as compared to 0.95g/cm^3 (0.551oz/cu.in.) for PS then the shot capacity needs to be down rated by 22%. What this means is that for a screw and barrel assembly having a shot capacity of 190g in PS, the actual shot capacity in HDPE will be 150g.

20 Melt Temperature

This is as measured in the nozzle or by an air shot technique; 205°C to 280°C (401°F to 536°F) is usually suggested. However, if melt temperatures are too low then there is a danger of high residual stresses and subsequent environmental stress cracking; the use of higher temperatures can result in thermal decomposition and so a better temperature range is 220 to 260°C (428°F to 500°F).

For high molecular weight material, temperatures of 200 to 250°C, that is 392 to 482°F, are suggested and with mold temperatures of 40 to 70°C, that is 104 to 158°F.

21 Barrel Residence Time

At a temperature of 270°C (518°F) the maximum residence time in the barrel should not exceed 5 to 6 minutes. When the temperature is 280°C (536°F) then the maximum residence time in the barrel should not exceed 2 to 3 minutes.

22 Temperature Settings

Please note that it is the melt temperature which is important; those temperatures given in the table are only suggested, initial settings. The temperature of the hydraulic oil, the hopper block and the material temperature should also not vary excessively. Mold temperature control is extremely important and therefore these temperatures must be checked regularly.

Zone No.	Location	Temperature in °C FROM	TO	Temperature in °F FROM	TO
0	Hopper throat	50	80	122	176
1	Barrel rear	160	200	320	392
2	Barrel middle	170	230	338	446
3	Barrel middle	200	260	392	500
4	Barrel front	220	280	428	536
5	Nozzle	210	270	410	518
6	Mold	10	60	50	140

23 Injection Speed

Moldings requiring a high surface gloss and good mechanical properties are molded using fast injection speeds. Fast injection speeds are also preferable for moldings where two or more melt fronts meet (i.e. converge) so as to ensure that good fusion takes place. When using fast injection speeds, good mold venting is essential to prevent burn marks occurring on molded surfaces. Profiled injection speeds are also used to overcome warpage problems associated with large surface area components, such as battery cases.

24 Injection Pressure

The machine should be capable of giving up to:

first stage, up to 1,500bar; 150MNm^{-2}; 21,750psi.
second stage, up to 750bar; 75MNm^{-2}; 10,850psi.

25 Screw Rotational Speed

Adjust speed (the rpm) to suit the molding cycle but do not use a screw surface speed of greater than 0.75 to 0.8ms^{-1} (2.46 to 2.62ft/sec). If a screw of 40mm (1.57in) diameter is being used then the rpm should not exceed 380. Similarly the maximum diameter of screw when rotated at 200rpm, in order to obtain a screw surface speed of 0.75 to 0.8ms^{-1} (2.46 to 2.62ft/sec), is 76mm (2.99in).

Some fast cycling grades of HDPE are processed using a screw surface speed of 0.95 to 1ms^{-1} (3.11 to 3.28ft/sec). These materials are specifically used for the molding of products related to the cosmetic and packaging industries.

26 Back Pressure

Should be kept as low as possible: usually a nominal value of 5 bar (0.5MNm^{-2}, 72.5psi) is used so as to achieve consistent shot weights. May need to be increased, to assist dispersion, when using colorant but keep as low as possible.

However, up to 150bar; 15MNm^{-2}; 2175psi, should be available.

27 Shutting Down

No need to purge the barrel with another material. Usual procedure is to purge barrel clean, leave the screw in its forward position and switch off the heats.

28 Reprocessing

When this material is reclaimed it is suggested that up to 50% regrind may be blended with the virgin material. In various instances 100% regrind can be used, but this is solely dependent upon the product and end-usage requirements. One of the main problems when using regrind material is the inability to maintain a consistent rate of material feed into the barrel assembly. Regranulated polyethylene is light and fluffy by nature and because of this, it has a tendency to bridge in the hopper thus creating feeding problems. More often than not virgin material is added to regrind to add weight (bulk) to the regrind/virgin mix so as to overcome the bridging problems. An alternative method is to use a vibratory feeding device mounted within the hopper itself.

29 Finishing

This material may not be joined to itself using solvents as there is no solvent at room temperature. Because of its inert, 'non-stick' surface it also cannot be very successfully bonded using adhesives; limited success with contact or hot melt adhesives. If the surface is made polar, for example by using a flame or an electrical discharge, then this material may be bonded to metals using epoxides or nitrile-phenolic adhesives; such treatments also improve printability. Flaming is popular; to ensure satisfactory adhesion the surface tension, after treatment, should be >44mN/m. this is achieved by using an oxidizing flame at temperatures of >1000°C (>1832°F): contact times are short - approximately a few hundredths of a second.

Because of HDPE's excellent resistance to solvents the use of solvent based adhesives is limited and therefore other forms of jointing methods such as snap fits, the use of self tapping screws and welding methods (such as hot plate or shoe) are often preferred. When welding HDPE, it is usual to coat or cover the hot plates with PTFE so as to prevent the material sticking to the surfaces of the hot plate. Components can be decorated by printing and surface coating, hot foiling or vacuum metallizing.

Machining of this plastic can be difficult because of its soft, resilient nature. Do not apply too much pressure when machining as the material will distort. HDPE can however, be readily machined with cutting tools used for metals; use high cutting speeds and low feed rates. The edges of the cutting tools must be sharp in order to achieve a good surface finish; use plenty of cutting fluid to cool the cutting surface and prevent any heat build up when machining. Because of it's low surface hardness it is extremely difficult to achieve a high gloss finish (comparable to a molded finish) on the machined surface.

30 Other Comments

The effects of changes in molecular weight, density and molecular weight distribution (MWD) are as follows.

Property.	Increase in the following will cause:		
	MFR	Density	Increase in breadth of MWD.
Pseudoplasticity.			Increases.
Tensile strength at yield		Increases.	
Tensile strength at break.	Decreases	Increases.	
Elongation at break.	Decreases	Decreases	
Impact strength.	Decreases	Slight decrease	Decreases
Modulus		Increases.	
Transparency		Decreases	
Long term load bearing		Increases.	
ESC resistance	Decreases	Decreases	
Softening temperature.		Increases.	
Melt strength.	Decreases		Increases.
Melt elasticity	Decreases		Increases.
Melt fracture tendency	Decreases		Decreases.
Gloss			Decreases.

HDPE may be cross-linked using high energy radiation or, by the incorporation of peroxides; carbon black reinforces such materials and also improves light resistance.

31 Typical Components

This material is strong and stiff, even at low temperatures. For these reasons it is used to produce components which have a reasonable impact strength at low temperatures, e.g bottle crates, containers used in refrigeration, fish boxes, storage vessels for paint, adhesives etc.. It is also used for containers where its rigidity and resistance to gas permeability is an asset when used for food storage boxes. Widely used for household goods e.g. buckets, bowls, kitchenware; although it has a higher density than PP the cost difference in £/kilo or $/lb still enables HDPE to compete with PP for various applications. Other uses for HDPE include dustbins, overcaps for the aerosol spray cans, security seal caps for medicine bottles, bases for PET bottles (i.e. fizzy drink bottles). and toys.

A major use for HDPE is in blow molding applications. Large containers or drums (for example, of 180 litres/45 gallons capacity) toys and car components (fuel tanks and spoilers) are typical applications. HDPE is used in such applications because of its high impact strength and, because it can retain its properties at low temperatures, for example, $-18°C$ ($-0.4°F$). To obtain the required properties in the finished product (for example, strength and creep resistance), it is often necessary to use high molecular weight material - known as HMW HDPE or, as HDPE HMW.

For blow molding, most commercial HDPEs have molecular weights in the range of 50,000 to 250,000. HMW material will be higher than this, say 250,000 to 1 million. Ultra high molecular weight materials (known as UHMW HDPE or HDPE-UHMW) will have molecular weights greater than 1 million; it has tremendous properties but molding is very difficult. Now, a high molecular weight HDPE is available for injection molding (Hostalen GUR GX). It combines the characteristically high abrasion resistance, low friction and excellent low temperatures, impact strength of HMW material with adequate flow properties for the volume production of large and complex components. Uses will include machinery components, sports goods and medical items, for example, orthopedic items.

Plastics materials can be made biodegradable, for example, by the use of additives or, by making special polymers. By copolymerization of ethylene with carbon monoxide, it is possible to produce a material which will be biodegradable: as the carbon monoxide level increase so does the brittleness. This material can be made more biodegradable than paper board.

32 Unit Price

The unit price compared to GPPS (based on one tonne lots) is:

Easy flow grade	0.82
General purpose or medium flow grade	0.82
Stiff flow grade	0.86.

The price relative to GPPS can also be found in Table 7.

LOW DENSITY POLYETHYLENE.

1 Common Name

Low density polyethylene.

2 Abbreviation(s)

LDPE ; PE-LD .

3 Alternative Names

Polythene; low density polythene; polyethene; poly(methylene); high pressure polyethylene.

4 Some Suppliers

Asia Polymer Co
Atochem
Bayer
BASF
BP
Chemie Linz
Dow
DuPont
DSM
Eastman Chemicals
Enimont
Enimont
Exxon
Neste
Novacor
Orkem
Petroleos Mexicanos
Petroquimica Triunfo
Shell
Statoil
USI

5 Trade Names or Trade Marks

Polymer E
Laqtene
Baylon
Lupolen
Novex LD
Daplen LDPE
Dowlex LDPE
Alathon
Stamylan LD
Tenite polyethylene
Eraclene
Fertene
Escorene
Neste LD
Novapol LD
Lotrene
Pemex
LDPE
Carlona PE
Statoil Polyethylene LDPE
Petrothene

6 Material Properties

When ethylene is heated at high pressures (up to 3,000 bar) and temperatures of 100 to 300°C (212°F to 572°F), in the presence of a free radical initiator, the plastic known as low density polyethylene is produced. This is a semi-crystalline, thermoplastics material and as such is not available as transparent moldings. Its natural color is a milky white and it has a soft, wax-like feel. Because of chain branching the crystallinity level is low and the material has a relatively low density, for example, 0.92 g.cm^{-3} (0.53 oz/cu.in). By varying the polymerization conditions it is possible to produce commercial materials with densities over the range 0.915 to 0.94g.cm^{-3} (0.53 to 0.55oz/cu.in). Both long chain and short chain branching is present and, because long chain, branched molecules are compact, the material flows relatively easily. This material is tough but only has moderate tensile strength and suffers from creep; for design purposes, 5% strain may be used as the operating limit. The impact resistance, chemical resistance and electrical insulation properties are excellent.

The water vapor permeability of this type of PE, as with all types of PE, is low. PE is permeable to gases and vapors (i.e. it lets them through) but LLDPE and HDPE are less permeable to gases and vapors than LDPE. Permeability for organic vapors is least for alcohols and then increases in the order shown; from acids to aldehydes and ketones, esters, ethers, hydrocarbons and halogenated hydrocarbons. (Permeability decreases with density).

When the density is 0.910 to 0.925g.cm^{-3} (0.53 to 0.54oz/cu.in), the material is sometimes known as Type 1; when the density is 0.926 to 0.94g.cm^{-3} (0.54 to 0.55oz/cu.in), it is sometimes referred to as Type 2. HDPE, with a density of 0.940 to 0.965g.cm^{-3} (0.55 to 0.56oz/cu.in), is sometimes known as Type 3.

By copolymerization with alpha olefins it is possible to produce a range of very low density materials (VLDPE.) which can have a density of 0.88 to 0.91g.cm^{-3} (0.51 to 0.53oz/cu.in.). Can be used as an alternative to thermoplastic elastomers and for materials modification, for example, for improving the impact strength of PP. The higher density materials are also used as injection molding plastics.

7 Ease of Flow

LDPE is an easy flow material. Ease of flow is rated by melt flow index (MFI) or rate (MFR); the lower the number, the stiffer is the flow as the molecular weight is higher. At 200°C (392°F) a plastic with an MFR of 20, would have approximately double the spiral flow length of a plastic with an MFR of 2. Low MFR materials exhibit better environmental stress cracking resistance (ESC), solvent resistance and higher impact strength. Polymers of different density but with the same melt flow index do not have the same molecular weight. More pseudoplastic than LLDPE. because it has a broader molecular weight distribution. If a LDPE plastic of a certain melt flow rate (index), is compared with an LLDPE. plastic of the same melt flow index, then it will be found that the linear polymer will be of higher viscosity at processing shear rates.

A close look at the values given in the Moldflow table, shows that at low shear rates the easy flowing grade has a higher viscosity than the medium flow grade; whereas, at the higher shear rates encountered in injection molding the flow behaviour is similar to that of other thermoplastics materials.

MOLDFLOW DATA FOR BASF LUPOLEN LDPE

Material.	Viscosity (Nsm^{-2}) at 1,000s^{-1} at the following temperatures; in °C (°F)			
	200°C (392°F)	220°C (428°F)	240°C (464°F)	260°C (500°F)
Easy flow grade (2410T)	53	39	30	22
Medium flow grade (1800S)	48	35	25	18
Stiff flow grade (3020K)	113	91	73	59

Material Grade	Viscosity (Nsm^{-2}) at 220°C (428°F)			
	100s^{-1}	1,000s^{-1}	10,000s^{-1}	100,000s^{-1}
Lupolen 2410T	95	39	14	6
Lupolen 1800S	124	35	16	7
Lupolen 3020K	338	91	25	7

8 Shrinkage

This is of the order of 0.02 to 0.05mm/mm or in/in (ie 2 to 5%) when the density is 0.910 to 0.925g.cm^{-3} (0.53 to 0.54oz/cu.in). It is of the order of 0.015 to 0.04mm/mm or in/in (ie 1.5 to 4%) when the density is 0.926 to 0.94g.cm^{-3} (0.54 to 0.55oz/cu.in). Because of LDPE's wide shrinkage range it is extremely difficult to accurately predict the necessary shrinkage value in order to achieve the desired dimensions for a particular component. Because of this a sample cavity is manufactured and moldings produced using typical processing conditions (i.e. temperatures, pressures and cycle time). Shrinkage values are obtained from these moldings which are then used for the sizing of the cavity and core dimensions of the production mold.

9 Resistant to the Following

Most solvents at room temperature but aromatic and chlorinated hydrocarbons will cause swelling. Relatively unaffected by polar solvents ,for example alcohols, phenols, esters and ketones; vegetable oils, water, alkalis, most concentrated acids at room temperature and ozone (in absence of U.V.). Very low water absorption even after long immersion times, for example after one year at 20°C (68°F), the increase in weight may be less than 0.2%. The addition of carbon black, used to improve weathering, will increase the water absorption. Absorption of other liquids, for example acetone and benzene, will be greater for LDPE than for HDPE. Best chemical resistance is found with HDPE and cross-linked PE.

10 Not Resistant To

Fuming nitric acid and also fuming sulfuric acid; slowly attacked by halogens and chlorinating agents such as chlorosulfonic acid and phosgene. (Fluorine, a halogen, has been used to reduce the permeability of blow molded containers). LDPE is associated with environmental stress cracking; this occurs when a material is subject to strain, or an internal stress, in the presence of polar liquids or the vapors of such liquids. Such environmental stress cracking is also associated with detergents or silicone fluids although there are many other environmental stress cracking agents, for example, chloroform, xylene and paraffin. The light, or U.V., resistance is poor; the cheapest way to improve this is by incorporation of carbon black. Has limited resistance to oxygen at elevated temperatures, antioxidants are used for protection; unless so protected the electrical properties will suffer.

11 Material Detection or Identification

With a density of approximately 0.92gcm^{-3} the material (solid, non-filled material) will float in both water and in saturated magnesium chloride. The manufacturing processes used to produce PE are as follows.

Process	*Product*	*Density (kg/m^3)*
Gas phase	LLDPE, HDPE	900 to 960
High pressure	LDPE	915 to 935
Modified high pressure	LLDPE, HDPE	880 to 940
Slurry	HDPE	935 to 960
Solution	LLDPE, HDPE	900 to 960.

(If a material has a density of $900 kg/m^3$, then the SG will be 0.9)

No solvent at room temperature but at higher temperatures (approximately 55°C or 131°F) soluble in hydrocarbons and chlorinated hydrocarbons, for example xylene and trichloroethylene. Below 60°C (140°F) approximately, PE is insoluble in all organic solvents but it does swell in aliphatic, aromatic and chlorinated hydrocarbons: the lower the density the more it swells.

The natural color of the material is a milky white and so a wide color range is possible. When this material is heated in a flame it ignites easily and burns with a yellow-tipped, blue flame giving off only a little smoke; forms burning drops and when the flame is extinguished, gives a smell like candle grease. LDPE has a melting point of 110 to 125°C (230°F to 257°F), LLDPE 115 to 128°C (239°F to 262°F), HDPE is 130 to 135°C (266°F to 275°F) and PP is 165 to 175°C (329°F to 347°F).

When heated in the absence of a flame the material will soften and melt to give a clear liquid, as the crystal structures are destroyed. Is stable in the absence of air up to approximately 300°C (572°F) when it decomposes to give low molecular weight hydrocarbons. (Cross-linked PE will not melt but will become rubbery at approximately 115°C (239°F)). Can be cut easily with a knife and can also be easily scratched with a knife or fingernail.

12 Coloring.

As the natural color of the material is an off-white then a wide color range is possible; this does not include transparent colors. Sold in both compounded colors and as natural material for coloring on the injection molding machine by techniques such as dry coloring, masterbatching and liquid coloring.

When coloring LDPE, organic dyes should not be used due to the problem of color leeching or bleeding. For most coloring purposes the use of organic and inorganic pigments are preferred. When dry coloring LDPE colorant levels of up to and including 1% are generally used. Wetting agents are used with dry colorants primarily for cleanliness in the molding shop rather than aiding the dispersion of the colorant. Universal type masterbatches are often used with LDPE at concentration levels of 1%, however to obtain a more uniform color on the molded component the addition level may need to increased to 3 to 5%. When color uniformity is particularly important LDPE based masterbatches are often preferred. If opacity of color is required then inorganic pigments tend to be used. There are no real limitations to using liquid colorants with LDPE, however colorant addition levels have to be kept to a minimum (i.e. no more than 3%) otherwise screw slip can occur.

More often than not molders use the same approach as for coloring GPPS, however the significant difference between PS and LDPE is the softness of LDPE and its very easy flow characteristics. To improve the dispersion of the pigments within the melt it is necessary to create a adequate mixing/grinding action within the barrel of the machine. In order to achieve this it is often necessary to reduce the temperatures on the machine to the lowest value that will still enable components to be produced of the required quality.

13 Materials and Components Handling

Will absorb less than 0.02% water in 24 hours at room temperature. This means that drying is not normally necessary. If it is, dry in a hot air oven for 3 hours at 65°C (149°F), in a desiccant dryer for 1 to 1½ hours at 80°C (176°F).

This material may not be joined to itself using solvents as there is no solvent at room temperature. Because of its inert, 'non-stick' surface it also cannot be very successfully bonded using adhesives; limited success with contact or hot melt adhesives. If the surface is made polar, for example by using a flame or an electrical discharge, then this material may be bonded to metals using epoxies or nitrile-phenolic adhesives; such treatments also improve printability. Commonly welded using techniques such as hot plate or hot shoe. Machining of this plastic is difficult because of its soft, non-resilient nature. Do not apply too much pressure when machining as the material will distort and tear quite easily.

14 Mold and Gate Considerations

A suggested mold temperature is 30°C (86°F). If a melt temperature of 210°C (410°F) is used then the maximum amount of heat that needs to be removed for each gram of shot is 572 Joules (J). Over the temperature range quoted, the specific heat is approximately 3180 J/kgK.

Typical runner sizes range from 4mm to 7mm (0.157 to 0.276in) diameter. Diameters above 7mm (0.276in) tend only to increase the cycle time and do not significantly assist in the filling of the mold. Runner lengths should be kept as short as possible and suitably sized ejectors positioned at runner junctions to ensure that the feed system is removed from the mold without difficulty.

All types of gates are used for LDPE, however the main criteria for each type of gate is its ability to be removed from the molded component without leaving unsightly gate scars or any vestige. As a result of the ductile or elastic nature of LDPE, most types of gates are usually removed from the component by a separate degating operation. Typical degating methods used are cropping, cutting by means of scissors or scalpel, or by simple twisting. Gate lands for the edge, pin or film type gates should not be greater than 0.75mm (0.030in) in length. Submarine type gates and reverse tapered sprue pullers are widely used for cold runner molds intended to run fully automatically. To ensure that minimal gate vestige occurs, when using a submarine gate, an inclusive gate angle of 30 degrees and a correctly positioned ejector pin (i.e. not greater than 10mm (0.394in) from gate entry) are essential.

LDPE is ideally suited for runnerless type molds due to its wide processing range and good thermal stability. All types of runnerless systems can be used for LDPE; however for production runs that require frequent color changes, the insulated hot tip type runnerless system is preferred. Typical gate diameters used for runnerless molds are 0.75 to 0.8mm (0.030 to 0.032in), however gate diameters as low as 0.6mm (0.024in) are often used to prevent gate vestige occurring - particularly when producing cosmetic components that require zero gate vestige. An alternative method to prevent gate vestige, or a short fiber attached to the gate, is to use hot runner nozzles that incorporate a shut off valve or needle. The incorporation of the valve prevents any plastics material being present in the gate aperture during degating. However the cost of these devices sometimes outweigh their effectiveness, particularly on large multi-impression molds.

Accurate temperature control of the runner manifold and nozzle assembly is necessary, particularly when using melt temperatures of 240°C (464°F) and above, so as to produce moldings of a consistent color. Poor temperature control can often lead to degradation of the pigment which results in color changes occurring during the production run.

If cold runner molds are used in preference to runnerless molds, for economic reasons, then hot sprue bushings should be incorporated in the mold to reduce the material content within the feed system and to allow faster cycle times to be achieved. Particular attention should be given to the design of the cooling circuitry in the mold when molding LDPE. To prevent large shrinkage variations and/or warpage occurring to moldings of large surface areas, or differing wall sections, the need for separate cooling circuits in the same half of the mold may be required. Each circuit should be independently controlled at a specific temperature, for example, in the region of the gate a mold surface temperature of between 10 to 20°C (50 to 68°F) would be used, whereas a temperature of between 25 to 60°C (77 to 140°F) would be used at the end of the melt flow path. By using different mold surface temperatures the overall cooling rate of the molding is much more uniform giving better dimensional stability. The position of the water channel in relation to the mold surface is very important when designing molds to be used with LDPE. Because of the large amounts of heat that need to be removed from the mold in order to achieve economic cycle times the diameter of the water cooling channel should not be less than 8mm (0.315in) and the distance from the mold surface to the edge of the cooling channel should not be greater than

1.5 times the cooling channel diameter.

Regranulation of the feed system of a cold runner type mold, may be accomplished by using a work handling device so that the feed system, when removed from the mold, is fed directly into the regranulator. Use of this technique ensures that minimal contamination occurs during the regranulation operation.

15 Flow Path:Wall Thickness ratio

This material is a very easy flow material; more easy flowing than PP or PS. At 1mm (0.039in) wall thickness the flow path: wall thickness ratio can reach 300:1 for a high MFR grade.

16 Projected Area.

Usually 1 to 2tsi. (15 to 30MNm^{-2}), is sufficient.

17 Cylinder Equipment

General purpose screws can be used ,for example, with an L:D ratio of about 20:1 and a constant pitch of 1D. The feed zone should be approximately 0.5L, the compression zone 0.3L and the metering zone 0.2L; a compression ratio of about 2:1 should be sufficient. The screw is usually fitted with a non-return valve but valves are not essential on the nozzle unless high back pressures are involved. The best results have been obtained by spring loaded or hydraulically operated types; decompression or suck back is also used.

18 Screw Cushion

On smaller machines this is approximately 3mm (0.118in); on larger machines 9mm (0.354in) is suggested.

19 Shot Capacity

Because of the materials good heat stability as little as 10% of the cylinders rated capacity may be used: the maximum possible is 95%. As the melt density of this material 0.75g/cm^3 (0.435oz/cu.in.), compared to 0.95g/cm^3 (0.551oz/cu.in.) for PS, the shot capacity for the screw and barrel assembly should be downrated by 22% e.g. if the shot capacity for PS is 145g, then the actual shot capacity for LDPE is 114g.

20 Melt Temperature

This is as measured in the nozzle or by an air shot technique;

180°C to 280°C (356°F to 536°F) depending upon grade and MFR.

21 Barrel Residence Time.

At a temperature of 270°C (518°F) the maximum residence time in the barrel should not exceed 5 to 6 minutes. When the temperature is 285°C (545°F) then the maximum residence time in the barrel should not exceed 2 to 3 minutes. If LDPE is overheated in the barrel, oxidation will occur giving rise to a reduction in the electrical insulation properties of the molded component.

22 Temperature Settings

Zone No.	Location	Temperature °C FROM	TO	Temperature °F FROM	TO
0	Hopper throat	40	70	104	158
1	Barrel rear	120	200	248	392
2	Barrel middle	160	230	320	446
3	Barrel middle	180	260	356	500
4	Barrel front	200	280	392	536
5	Nozzle	210	270	410	518
6	Mold	10	60	50	140

Please note that it is the melt temperature which is important; those temperatures given in the table are only suggested, initial settings. The temperature of the hydraulic oil, the hopper block and the material temperature should also not vary excessively. Mold temperature control is extremely important and therefore these temperatures must be checked regularly.

23 Injection Speed

Moldings requiring a high surface gloss and good mechanical properties are molded using fast injection speeds. Fast injection speeds are also preferable for moldings where two or more melt fronts meet (i.e. converge) so as to ensure that good fusion takes place. When using fast injection speeds, good mold venting is essential to prevent burn marks occurring on molded surfaces. Profiled injection speeds are also used to overcome warpage problems associated with large surface area components.

24 Injection Pressure

The machine should be capable of giving up to:

first stage, up to 1500bar; 150MNm^{-2}; 21,750psi.
second stage, up to 750bar; 75MNm^{-2}; 10,850psi.

Careful control over dwell time and pressure is needed in order to avoid overpacking but yet still be able to compensate for shrinkage.

25 Screw Rotational Speed

Adjust speed (the rpm) to suit the molding cycle but do not use a screw surface speed of greater than 0.7 to 0.75ms^{-1} (2.30 to 2.46ft/sec) If a screw of 40mm (1.57in) diameter is being used then the rpm should not exceed 360. Similarly the maximum diameter of screw when rotated at 200rpm in order to obtain a screw surface speed of 0.7 to 0.75ms^{-1} (2.30 to 2.46ft/sec) is 70mm (2.76in).

26 Back Pressure.

Up to 150bar; 15MNm^{-2}; 2,175psi. Should be kept as low as possible: usually a nominal value of 5 bar (0.5MNm^{-2}, 72.5psi) is used so as to achieve consistent shot weights. May need to be increased, to assist dispersion, when using colorant but keep as low as possible.

27 Shutting Down

No need to purge with another material.

28 Reprocessing

When this material is reclaimed it is suggested that up to 50% regrind may be blended with the virgin material. In various instances 100% regrind can be used, but this is solely dependent upon the product and end usage requirements. One of the main problems when using regrind material is the inability to maintain a consistent rate of material feed into the barrel assembly. Regranulated polyethylene is light and fluffy by nature and because of this it has a tendency to bridge in the hopper creating feeding problems. More often than not virgin material is added to regrind to add weight (bulk) to the regrind/virgin mix so as to overcome the bridging problems. When regrind material is added to virgin it is essential that the regrind used should be of the same density and/or MFR as the virgin polymer otherwise a deterioration in the physical properties of the resultant material will occur.

29 Finishing

The surface of this material may be made more receptive to inks or adhesives by pre-treatment using a corona discharge or by ozone treatment. This material is difficult to deflash because of the flexible nature of the material.

30 Other Comments

May be cross-linked using high energy radiation or by the incorporation of peroxides. Carbon black reinforces such materials and also improves light resistance.

31 Typical Components

Because of this material's ease of molding and low cost it has become established as a general purpose injection molding material. Most of the applications do not utilize the excellent electrical insulation properties, nor the water resistance of the material. LDPE is used for example for caps and lids for containers, bottle closures and tear-off closures, bowls, beakers, pipe couplings, pots, linen baskets, bins etc.. Close tolerances are difficult to mold in this material and they are difficult to hold in service because of the high coefficient of thermal expansion and the tendency to creep. The softness of this material can allow attack by insects or animals.

32 Unit Price

The unit price compared to GPPS (based on one tonne lots) is approximately:

Easy flow grade	0.82
General purpose or medium flow grade	0.82
Stiff flow grade	0.84

The price relative to GPPS can also be found in Table 7.

LINEAR LOW DENSITY POLYETHYLENE.

1 Common Name

Linear low density polyethylene.

2 Abbreviation(s)

LLDPE ; PE-LLD.

3 Alternative Names

Linear low: linear low density polythene.

4 Some Suppliers.   5 Trade Names or Trade Marks

Some Suppliers	Trade Names or Trade Marks
Atochem	Lacqtene HX
BP	Innovex
Dow Chemicals	Dowlex
DSM	Stamylex
DuPont	Sclair
Enron/Delong Prochem	Norchem
Exxon	Escorene
Neste	Neste LPLD
Northen Petrochemicals	Norlin
Orkem	Lotrex
USI	Petrothene

6 Material Properties

This material was developed in the 1950's but it is only within the last few years that it has come into prominence. This plastic is, in fact, a copolymer and is made by polymerizing ethylene with a small percentage of a higher olefin, for example, octene or butene. (Butene-type materials can be made by any of the present available processes and are cheaper; octene-type materials have slightly better mechanical properties. When these materials were first introduced, the lower price of them forced their use: the standard butene-based product was, in effect, being used as a filler for LDPE). With suitable catalysts (halogen-free metal complexes) only low pressures and temperatures are required for manufacture and polymerization efficiency is high. It is possible to produce PE over the density range 0.89 to $0.955 g/cm^3$ (0.52 to 0.55 oz/cu.in) and, it is also possible to produce materials which have a particular molecular weight: such materials have a narrow molecular weight distribution.

The comonomer introduces short chain branches into the

molecular structure (more than HDPE) and so it is possible to produce a material, which in some respects is similar to LDPE, but without the need for high polymerization pressures. These short chain branches interfere with chain packing and so, crystallization is not so pronounced as for HDPE; such materials have a lower density therefore than HDPE. (The term 'linear' means that there is an absence of long chain branching). Commercial materials have densities in the range 0.89 to 0.955g.cm^{-3} (0.52 to 0.55 oz/cu.in) and have a narrower molecular weight distribution than LDPE. They contain a smaller quantity of low molecular weight material and more medium molecular weight molecules; the average length of the molecular chain is longer. For a given molecular weight therefore, LLDPE will have better mechanical properties at low (e.g -40°C/-40°F) and high temperatures, better environmental stress cracking resistance, and higher melt viscosity than LDPE. It can be stronger and tougher than LDPE and because of it's narrow molecular weight distribution warp free moldings can be readily produced.

Increasing the comonomer content increases the flexibility, the impact and tear strength, the environmental stress cracking resistance and the ease of sealing and printing. However, the stiffness, yield and creep strength, the hardness, Vicat softening point and the melting point decrease proportionally. Decreasing the density, or the crystallinity, also increases the flexibility, the impact and tear strength, the environmental stress cracking resistance and the ease of sealing and printing. Similarly, the stiffness, yield and creep strength, the hardness, Vicat softening point and the melting point decrease proportionally.

Higher alpha olefin grades (HAO) of PE, are co- or terpolymers which contain up to 10% of octene, 4-tetramethylpentene-1 or, propylene. VLDPE, sometimes known as ULDPE, means very low density or, ultra low density and, such materials can have densities in the region of 0.88 to 0.91g.cm^{-3} (0.51 to 0.53 oz/cu.in). Such very light HAO grades are hardly crystalline and are rubber-like materials. They can be highly filled and used like other thermoplastics materials or, they can be used to enhance certain properties (for example, crack resistance) of other polyolefins.

7 Ease of Flow

Less pseudoplastic than LDPE; if a LDPE plastic of a certain melt flow rate (index), is compared with an LLDPE plastic of the same melt flow index, then it will be found that the linear polymer will be of higher viscosity at processing shear rates. MFR values can reach 100; with LDPE they can reach 250.

(Polymers of different density but with the same melt flow index do not have the same molecular weight). Replace a LDPE grade, with an MFR of 7 to 10, with a LLDPE of MFR of 20 to 50. No Moldflow data available.

8 Shrinkage

This is of the order of 0.015 to 0.030in/in or mm/mm ie 1.5 to 3.0%. By optimizing the molding conditions for LLDPE it is possible to obtain lower shrinkage values than those obtained with LDPE and HDPE. LLDPE also exhibits a much lower differential shrinkage than LDPE and so warping is considerably reduced.

9 Resistant to the Following

PE is associated with environmental stress cracking; this occurs when a material is subject to strain, or an internal stress, in the presence of polar liquids or the vapors of such liquids. Such environmental stress cracking is also associated with detergents or silicone fluids although there are many other environmental stress cracking agents, for example, chloroform, xylene and paraffin. Has better ESC resistance than LDPE. Higher molecular weight grades have the best ESC.
The water vapor permeability of this type of PE, as with all types of PE, is low. PE is permeable to gases and vapors (i.e. it lets them through) but LLDPE and HDPE are less permeable to gases and vapors than LDPE. Permeability for organic vapors is least for alcohols and then increases in the order shown; from acids to aldehydes and ketones, esters, ethers, hydrocarbons and halogenated hydrocarbons. (Permeability decreases with density). Some grades of LLDPE are accepted as being suitable for containers for oil and petrol: they have been used for fuel tanks: in some cases the formed containers do not have to be chemically modified so as to make the material almost impermeable to fuels.

10 Not Resistant To

The light, or U.V., resistance is relatively poor and is similar to that of LDPE. Cheapest way of improving this is by black incorporation, for example, addition of say 2% black can improve the weathering resistance by twelve times.. Has limited resistance to oxygen at elevated temperatures; antioxidants are used for protection. Unless so protected, the electrical properties will suffer.

11 Material Detection or Identification

With a density over the range 0.90 to 0.94gcm^{-3} (0.52 to 0.55 oz/cu.in) the material (solid, non-filled material) will float in both water and in saturated, magnesium chloride. The manufacturing processes used to produce PE are as follows.

Process	Product	Density (kg/m^3)
Gas phase	LLDPE, HDPE	900 to 960
High pressure	LDPE	915 to 935
Modified high pressure	LLDPE, HDPE	880 to 940
Slurry	HDPE	935 to 960
Solution	LLDPE, HDPE	900 to 960.

(If a material has a density of 900kg/m^3, then the SG will be 0.9)

There is no solvent at room temperature but at higher temperatures (approximately 55°C or 131°F) soluble in hydrocarbons and chlorinated hydrocarbons, for example xylene and trichloroethylene. Below 60°C (140°F) approximately, PE is insoluble in all organic solvents but it does swell in aliphatic, aromatic and chlorinated hydrocarbons: the lower the density the more it swells.

The natural color of the material is a translucent white. When this material is heated in a flame then it ignites easily, burns with a yellow-tipped, blue flame and gives off only a little smoke; forms burning drops and when the flame is extinguished, gives off a smell like candle grease. Can be distinguished from other polyolefins by density and melting point. LDPE has a melting point of 110 to 125°C (230 to 257°F), LLDPE 115 to 128°C (239 to 262°F), HDPE is 130 to 135°C (266 to 275°F) and PP is 165 to 1750C (329 to 347°F).

When heated in the absence of a flame then the material will soften and melt to give a clear liquid as the crystal structures are destroyed. Is stable in the absence of air up to approximately 300°C (572°F) when it decomposes to give low molecular weight hydrocarbons (Cross-linked PE (PE-V) will not melt but will become rubbery at approximately 115°C or 239°F). Can be cut easily with a knife and can also be easily scratched with a knife or fingernail.

12 Coloring

As the natural color of the material is an off-white then a wide color range is possible; this does not include transparent colors. Sold in both compounded colors and as natural material for coloring on the injection molding machine by techniques such as dry coloring, masterbatching and liquid coloring. When dry coloring, adhesion promoters such as paraffin can be used. Control the mixing time precisely as too short a mixing time gives poor dispersion and inadequate adhesion; too long a mixing time may cause pigment compaction. Pigment selection is important as some types of pigments increases the sensitivity for stress cracking and/or warpage to occur to moldings. When coloring LLDPE organic dyes should not be used due to the problem of color leeching or bleeding. For most coloring purposes the use of organic and inorganic pigments are preferred. Colorant levels of up to and including 1% are generally used. Adhesion promoters (i.e. wetting agents) are used with dry colorants primarily for cleanliness in the molding shop rather than aiding the dispersion of the colorant. Universal or LDPE based masterbatches are often used with LLDPE at concentration levels of 1%, however to obtain a more uniform color on the molded component the addition level may need to increased to 3 to 5%. When color uniformity and optimum end-use performance are particularly important, LLDPE based masterbatches are often preferred. If opacity of color is required, then inorganic pigments tend to be used. There are no real limitations to using liquid colorants with LLDPE, however colorant addition levels have to be kept to a minimum (i.e. no more than 3%) otherwise screw slip can occur. LLDPE's higher melt viscosity, compared to LDPE, enables better colorant dispersion even with the older types of molding machines. However, to improve the dispersion it is necessary to obtain a more efficient mixing/grinding action within the barrel of the machine. This can be achieved by reducing the temperatures on the machine to the lowest value that will still enable components to be produced of the required quality.

When using carbon black as a dry colorant care should be taken to prevent it from absorbing any moisture during storage otherwise the colored polymer will need to be dried before molding.

13 Materials and Components Handling

LLDPE will absorb less than 0.02% water in 24 hours at room temperature. This means that drying is not normally necessary. If it is, dry in a hot air oven for 3 hours at 65°C (149°F), in a desiccant dryer for 1 to $1^1/2$ hours at 85°C (185°F).

This material may not be joined to itself using solvents as there is no solvent at room temperature. Because of its inert, 'non-stick' surface it also cannot be very successfully bonded using adhesives; some success with contact or hot melt adhesives. If the surface is made polar, for example by using a flame or an electrical discharge, then this material may be bonded to metals using epoxies or nitrile-phenolic adhesives; such treatments also improve printability.

14 Mold and Gate Considerations

A suggested mold temperature is 40°C (86°F). If a melt temperature of 210°C (446°F) is used then the maximum amount of heat that needs to be removed for each gram of shot is 545 Joules(J). Over the temperature range quoted, the specific heat is approximately 3206 J/kgK.

Because of the rigidity of this material, compared to LDPE, moldings can be made of thinner wall thickness; the higher crystallization temperature means a faster set-up time. Both considerations should result in significantly faster cycle times.

Typical runner sizes range from 4mm to 7mm (0.157 to 0.276in) diameter. The runner diameter should be sufficiently large to prevent premature crystallization taking place thus reducing the effective runner diameter. However diameters above 7mm (0.276in) tend only to increase the cycle time and do not significantly assist in the filling of the mold. Runner lengths should be kept as short as possible and suitably sized ejectors positioned at runner junctions to ensure that the feed system is removed from the mold without difficulty.

All types of gates are used for LLDPE, however the main criteria for each type of gate is its ability to be removed from the molded component without leaving unsightly gate scars or any vestige. Typical degating methods used are cropping, cutting by means of scissors or scalpel, or by simple twisting. Gate lands for the edge, pin or film type gates should not be greater than 0.75mm (0.030in) in length. Submarine type gates and reverse tapered sprue pullers are widely used for cold runner molds intended to run fully automatically. To ensure that minimal gate vestige occurs, when using a submarine gate, an inclusive gate angle of 30 degrees and a correctly positioned ejector pin (i.e. not greater than 10mm (0.394in) from gate entry) are essential.

LLDPE is ideally suited for runnerless type molds due to its wide processing range and good thermal stability. All types of runnerless systems can be used for LLDPE; however, for production runs, that require frequent color changes, the insulated hot-tip type runnerless system is preferred. The use of heated probes in the runnerless system ensures that the temperature of the material in the gate region is not considerably lower than in the barrel nozzle thus, preventing premature crystallization taking place. Typical gate diameters used for runnerless molds are 0.75 to 0.8mm (0.030 to 0.032in), however gate diameters as low as 0.7mm (0.028in) are often used to prevent gate vestige occurring particularly when producing components that require zero gate vestige. To prevent gate vestige, or a short fiber attached to the gate, it is best to use hot runner nozzles that incorporate a shut-off valve or needle.

Accurate temperature control of the runner manifold and nozzle/probe assembly is necessary, particularly when using melt temperatures of 240°C (464°F) and above, so as to produce moldings of a consistent color. Poor temperature control can often lead to degradation of the pigment which results in color changes occurring during the production run.

Because LLDPE exhibits lower differential shrinkage values the problem of component warpage is considerably reduced compared to HDPE and LDPE. As a result of this the design of the mold cooling circuitry for LLDPE is not as critical as it is for LDPE and HDPE. The increased stiffness of LLDPE also allows the molding to be ejected from the mold at elevated temperatures without the risk of distortion.

When producing moldings of large surface areas or thick wall sections the need for separate cooling circuits in the same half of the mold may be necessary. Each circuit should be independently controlled at a specific temperature for example, in the region of the gate a mold surface temperature of between 10 to 20°C (50 to 68°F) would be used, whereas a temperature of between 25 to 60°C (77 to 140°F) would be used at the end of the melt flow path. Use of different mold surface temperatures ensures that the overall cooling rate of the molding is more uniform minimizing component warpage and improving dimensional stability.

The diameter of the water cooling channel should not be less than 8mm (0.315in) and the distance from the mold surface to the edge of the cooling channel should not be greater than 1.5 times the cooling channel diameter (1.5D).

For moldings that possess both thick and thin sections, the distance of the water channel should be closer for the thicker sections (i.e. 1.0 to 1.1D) than for the thinner sections. To achieve better cooling, particularly when molding thick sectioned components, certain areas of the mold should be made from beryllium copper in order to dissipate the heat more rapidly.

As fast injection speeds are used for LLDPE, burn marks or matt patches may appear on the molding, as a result of inadequate venting, so the position and size of the vents play an important part on the surface finish of the molded component. For example, the location of the vent could be at the end of the melt flow path, at the junction of melt flows or, at the base of deep (or blind) pockets. Vent sizes of 0.025 to 0.08mm (0.001 to 0.003in) deep by 4 to 6mm (0.157 to 0.236in) wide) are used.

Typical draft angles of 1 to 2 degrees per side are recommended. Use the maximum size ejector pins wherever possible or, if inconvenient consider using stripper sleeves, rings or plates so as to obtain the maximum amount of surface contact area to eject the component.

As LLDPE has good gloss characteristics, the use of highly polished mold surfaces, will produce moldings possessing surfaces of a very high gloss.

15 Flow Path:Wall Thickness Ratio

This material is a easy flowing; at 1mm(0.039in) wall thickness the flow path: wall thickness ratio can reach 250:1 for a high M.F.R. grade.

16 Projected Area

Usually 1 to 2tsi. (15 to 31MNm^{-2}) is sufficient. For components of thin wall sections and requiring a long flow length this may have to be increased to 2.5 to 3 tsi. (39 to 47MNm^{-2})

17 Cylinder Equipment

General purpose screws can be used ,for example, with an L:D ratio of about 20:1 and a constant pitch of 1D. The feed zone should be approximately 0.5L, the compression zone 0.3L and the metering zone 0.2L; a compression ratio of about 2:1 should be sufficient. The screw is usually fitted with a non-return valve but valves are not essential on the nozzle unless high back pressures are involved. The best results have been obtained by spring loaded or hydraulically operated types; decompression or suck back is also used.

18 Screw Cushion

On smaller machines this is approximately 3mm (0.118in); on larger machines 9mm (0.354in) is suggested.

19 Shot Capacity

Because of the materials good heat stability as little as 10% of the cylinders rated capacity may be used: the maximum possible is 95%. As the melt density of this material is 0.75g/cm^3 (0.435oz/cu.in.), compared to 0.95g/cm^3 (0.551oz/cu.in.) for PS, the shot capacity for the screw and barrel assembly should be downrated by 22% e.g. if the shot capacity for PS is 150g, then the actual shot capacity for LLDPE is 118g.

20 Melt Temperature

This is as measured in the nozzle or by an air shot technique; 160°C to 280°C (392 to 536°F) depending upon the density: the use of the higher melt temperatures can result in thermal degradation occurring and therefore the more suitable range is 160 to 240°C (356 to 464°F). Can be processed at melt temperatures as low as 140°C (284°F) when using very easy flow grades. The use of such temperatures enables LDPE to be replaced by LLDPE, particularly for thick sectioned components, in order to reduce the cycle time without experiencing problems of warpage and/or distortion.

21 Barrel Residence Time

At a temperature of 270°C (518°F) the maximum residence time in the barrel should not exceed 5 to 6 minutes. When the temperature is 285°C (545°F) then the maximum residence time in the barrel should not exceed 2 to 3 minutes

22 Temperature Settings

Zone: No.	Location	Temperature °C FROM	TO	Temperature °F FROM	TO
0	Hopper throat	40	70	104	158
1	Barrel rear	130	200	266	392
2	Barrel middle	140	230	284	446
3	Barrel middle	160	260	320	500
4	Barrel front	160	280	320	536
5	Nozzle	160	270	320	518
6	Mold	10	60	50	140

Please note that it is the melt temperature which is important; those temperatures given in the table are only suggested, initial settings. The temperature of the hydraulic oil, the hopper block and the material temperature should also not vary excessively. Mold temperature control is extremely important and therefore these temperatures must be checked regularly.

23 Injection Speed

Moldings requiring a high surface gloss and good mechanical properties are molded using fast injection speeds. Such speeds are used so that the mold is filled before the material flowing to the extremities has time to cool down. This ensures that uniform crystallization takes place which produces moldings having lower internal stresses. Fast injection speeds are also preferable for moldings where two or more melt fronts meet (i.e.converge) so as to ensure that good fusion takes place. When using speeds of a fast nature good mold venting is essential to prevent defects such as burn marks or matt areas appearing on molded surfaces.

24 Injection Pressure

The machine should be capable of giving up to:
first stage, up to 1500bar; 150MNm^{-2}; 21,750psi.
second stage,up to 750bar; 75MNm^{-2}; 10,850psi.

25 Screw Rotational Speed

Adjust speed (the rpm) to suit the molding cycle but do not use a screw surface speed of greater than 0.7 to 0.75ms^{-1} (2.30 to 2.46ft/sec). If a screw of 40mm (1.57in) diameter is being used then the rpm should not exceed 360. Similarly the maximum diameter of screw when rotated at 200rpm in order to obtain a screw surface speed of 0.7 to 0.75ms^{-1} (2.30 to 2.46ft/sec) is 70mm (2.76in).

26 Back Pressure

Up to 150bar; 15MNm^{-2}; 2,175psi should be available .Should be kept as low as possible: usually a nominal value of 5 bar (0.5MNm^{-2}, 72.5psi) is used so as to achieve consistent shot weights. May need to be increased, to assist dispersion, when using colorant but keep as low as possible.

27 Shutting Down

There is no need to purge with another material.

28 Reprocessing

When this material is reclaimed it is suggested that up to 50% regrind may be blended with the virgin material. In various instances 100% regrind can be used, but this is solely dependent upon the product and end usage requirements. One of the main problems when using regrind material is the inability to maintain a consistent rate of material feed into the barrel assembly. Regranulated polyethylene is light and fluffy by nature and because of this it has a tendency to bridge in the hopper creating feeding problems. More often than not virgin material is added to regrind to add weight (bulk) to the regrind/virgin mix so as to overcome the bridging problems.

When regrind material is added to virgin it is not as critical as with LDPE that the regrind should be of the same density and/or MFR as the virgin polymer.

Regranulation of the feed system of a cold runner type mold may be accomplished by using a work handling device or sprue grabber, so that when the feed system, when removed from the mold, is fed directly into the regranulator. Use of this technique ensures that minimal contamination occurs during the regranulation operation.

29 Finishing

The surface of this material may be made more receptive to inks or to adhesives by pre-treatment using a corona discharge or by ozone. This material is difficult to deflash because of it's flexible nature. Commonly welded using techniques such as hot plate or hot shoe; when welding LLDPE it is usual to cover the plates, or shoe, with PTFE so as to prevent the melted material sticking to the hot surfaces Machining of this plastic can be carried using conventional machining methods and cutting tools. High cutting speeds and low feed rates are necessary to produce accurate machined surfaces: use plenty of cutting fluid to cool the cutting surface and prevent any heat build-up when machining. Surface grinding of LLDPE is extremely difficult and can only be carried out using specially developed processes.

30 Other Comments

May be cross-linked using high energy radiation or, by the incorporation of peroxides; carbon black reinforces such materials and also improves light resistance. The compatibility of LLDPE and HDPE is good where use of such blends is now being considered for industrial paint containers.

31 Typical Components

Film is the major market for this material. Because of the rigidity of this material and its higher elongation at break, and puncture resistance, compared to LDPE, film extrusions can be made of thinner thickness, that is, the product may be down-gaged. Similar considerations should apply to blow molding. Some grades of LLDPE are accepted as being suitable for containers for oil and petrol. They have been used, for example, to blow mold fuel tanks: in some cases the formed containers do not have to be chemically modified - this is done to make the material almost impermeable to fuels.

Injection moldings can have high gloss, good impact

strength and good resistance to environmental stress cracking. As a result of a combination of environmental stress cracking (ESC), dimensional stability and good torque retention, screw caps and closures, particularly in the pharmaceutical field, are made from LLDPE.

Blends of LLDPE with other polyethylenes are widely used in other branches of the plastics industry, for example, in film products. Now blends of HDPE and LLDPE are being considered for industrial paint containers. The addition of some LLDPE to HMW-MDPE (see HDPE) improves the stress crack resistance as it slightly disrupts the crystalline structure of this semi-crystalline, thermoplastics material. For a given MFR and density, LLDPE has a higher Vicat softening point (10 to 15°C, (18 to 27°F)) and lower internal stresses than LDPE; this means that it can be used for components subjected to elevated temperatures. Because of the material's good low temperature properties, it may be utilized for containers used in cold storage rooms and for high quality houseware and toys (specifically snowsledges and snowmobiles). The addition of very low density PE (VLDPE) will give the material increased elasticity and, improve the low temperature impact strength.

It is becoming increasingly used for large, thin, circular and rectangular box lids – because of the flatness required (in order to decorate the lid) and, because of the cheapness of the product obtained, as a result of the thinner sections being used.

LLDPE is now being considered for applications such as reusable containers, e.g. once an ice-cream container is emptied it becomes a quality sandwich box. The reason for such use is it's wide ranging properties and ability to be cleansed in dishwashers

32 Unit Price

If the price of a natural, unmodified grade of GPPS is taken as one then the relative price of this molding material is approximately

General purpose grade   0.98

The relative price to GPPS can also be found in Table 7.

POLYPROPYLENE

1 Common Name

Polypropylene.

2 Abbreviation(s)

PP: PPR: PPN: PP:H: PP-K: PP-C: PP-HO: PP-CO. (Also seen without the hyphens).

3 Alternative Names

Polypro; polypropene; poly(propylene); polymethylethylene.

4 Some Suppliers	5 Trade Names or Trade Marks
Advanced Global Polymers	Adpro
Amoco	Amoco PP
Apryl	Apryl
Atochem	Lacqtene P
Bamberger Polymers	Bergaprop
B.A.S.F.	Novolen.
B.P	Napryl
Enron/Delong Prochem	Norchem
Huls	Vestolen P
Cabot Plastics	Cabelec (conductive PP)
Chemie Linz	Daplen PP
DSM	Stamylan P
Eastman	Tenite
El Paso	Rexene
Himont Moplen,	Profax
Hoechst/Hoechst Celanese	Hostalen PP
I.C.I.	Propathene
Laporte Industries	Eltex P
Monmouth Plastics	Empee PP
Neste	Neste PP
Norchem Inc	Nortuff
Petrofina	Finaprop
Phillips Petroleum Ltd.	Marlex
Schulman	Polyflan
Shell	Shell PP
Soltex	Fortilene
Solvay	Eltex P
Statoil	Statoil PP

6 Material Properties

By the use of Ziegler-Natta catalysts, PP (which is mainly isotactic PP) may be produced from the monomer propylene i.e. the long chain homopolymer (PP-H or PPH) formed, has sterically, ordered side groups (all the methyl groups lie on one side of the chain). PP-H like PE, is a linear, hydrocarbon plastic but is stiffer, harder and has a higher melting point. (Pure isotactic PP is not commercially made as a molding material - has a melting point greater that 170°C (338°F) and an SG of approximately 0.91). PP-H has high strength and stiffness but a low notched impact strength. Becomes very brittle at about 0°C (32°F) and for this reason block copolymers, with ethylene, are often preferred. (Both random copolymers, with approximately 1 to 4% ethylene, and block copolymers, which can contain higher ethylene contents, are made).

It is the ethylene-propylene block copolymers which are traditionally used in place of the homopolymers. Such block, copolymer materials may be known as polyallomers or, as propylene copolymers and referred to as either PP or, PP-K or, as PP-C or, as PP-B. The K simply means that it is a copolymer: K is used instead of C as, in some circles, C stands for chlorination. PP-B stands for block copolymer. PP may therefore refer to either a homopolymer or, to a copolymer (PP-H and/or PP-K and/or PP-B). Copolymers have a lower heat distortion temperature (HDT), less clarity, less gloss and, less rigidity but greater impact strength. As the proportion of ethylene increases the material becomes softer and tougher.

PP (that is, PP-H and PP-K) can have a Rockwell hardness (R scale) of approximately 90, a density of $0.9 g.cm^{-3}$, a Vicat softening point of about 150°C (302°F) and, a heat distortion temperature of approximately 100°C (212°F). (Random copolymers have lower values than this). Moldings can withstand boiling water and steam sterilization and, they do not suffer from environmental stress cracking (ESC) problems; the maximum service temperature is above 100°C (212°F), for example, 110°C (230°F). Electrical insulation properties are good: a high gloss, scratch-resistant surface is possible.

PP is extensively modified by the addition of glass fibers, mineral fillers, thermoplastic rubbers or a combination of these. For example, the rigidity, hardness and heat distortion temperature of PP can be markedly improved by the incorporation of talc. Rubber modification improves the low temperature, impact strength (of both PP-H and PP-K) with some reduction in stiffness.

PP continues to develop and to create new markets: this is because of the versatility of the material. New polymerization technology, allows the production of directly polymerized, soft grades which compete with elastomer (rubber) modified PP (RRPP). Valtec is a PP that has not been subject to the traditional heat history of melt compounded PP. The spherical particles (1 to 4mm, or 0.039 to 0.157 in) are surface coated after production and, they melt faster as they are not crystalline (as supplied). Random copolymers are being promoted, for example, for blow molding applications as they are clear materials with a high gloss and a Vicat softening point (VST) of approximately, 128°C (263°F) (20°C (36°F) lower than PP-H). At the moment they are slightly more expensive than traditional materials.

7 Ease of Flow

Injection molding materials can have a melt flow rate (MFR) (230°C/21.6 N ie 2.16kg.) of between 1 and 40. The low values refer to materials with a high impact strength and the high values refer to materials with very easy flow characteristics. Increasing the molecular weight with PP (decreasing the MFR) reduces the tensile strength and stiffness but increases the impact strength as, high molecular weight material finds it difficult to crystallize. More temperature and shear sensitive than PE. Higher viscosity grades give lower cavitation, or void problems. PP-K is tougher than PP-H of the same nominal MFR; such materials offer advantages for low temperature applications. Controlled rheology grades (CR) of PP are produced by 'cracking', or breaking, high molecular weight chains so as to give a material with a narrow molecular weight spread (distribution) and, which flows easily.

MOLDFLOW DATA FOR BASF NOVOLEN PP

Material	Viscosity (Nsm^{-2}) at $1,000s^{-1}$ at the following temperatures in °C (°F)			
	200°C (392°F)	220°C (428°F)	240°C (4640F)	260°C (500°F)
Easy flow grade (1300H)	204	174	148	126
Medium flow grade (1300E)	212	182	156	134
Stiff flow grade)1300L)	273	246	222	199

Material Grade	Viscosity (Nsm^{-2}) at 240°C (428°F)			
	$100s^{-1}$	$1,000s^{-1}$	$10,000s^{-1}$	$100,000s^{-1}$
Novolen (1300H)	641	148	34	7
Novolen (1300E)	886	156	28	5
Novolen (1300L)	881	222	59	14

8 Shrinkage

Being a semi-crystalline, thermoplastics material, PP has a relatively high shrinkage which is of the order of 0.018 in/in or mm/mm, ie, 1.8%; can reach 2.5% in thick sections. Gives more uniform shrinkage than HDPE; the difference in shrinkage, in different directions, is usually ignored as it should be small, for example, 0.2%. Approximately 85% of the shrinkage takes place in the first 24 hours, 98% in the first week and the remainder may take up to 3 months. If the moldings are placed in boiling water, then the shrinkage is complete in 1 hour. An excessive amount of delayed shrinkage indicates that the wrong molding conditions were used, for example, low mold and melt temperatures and/or high pressures.

Such conditions can also give rise to component warping; this is caused by excessive orientation (which in turn causes differential shrinkage) or, by uneven cooling. Test for warping by placing the molding in boiling water for 15 minutes; if this does not cause excessive distortion then the injection molding conditions are satisfactory.

Shrinkage uniformity can be improved by removing the high molecular weight fraction, for example, by the use of peroxides. However the incorporation of such peroxides is a specialized process and therefore cannot be carried during the processing stage (i.e within the barrel of the molding machine). Such treatments give easy flow materials with a reduced tendency to warp. (Grades which have such a narrow molecular weight distribution may be known as 'controlled rheology' (CR) materials).

The addition of 30% glass fiber (may be known as PPGF30) can reduce the shrinkage to approximately 0.7%. PPGF has its lowest shrinkage in the direction of flow.

9 Resistant to the Following

A wide range of common solvents and organic liquids. Outstanding resistance to hot water and to chemicals. Molding materials are dissolved only at elevated temperatures by, for example, aromatic hydrocarbons such as xylene. Relatively unaffected by aqueous solutions, including quite strong acids and alkalis. Because of swelling problems, at room temperature, not recommended for use with aromatic hydrocarbons (for example, benzene), and chlorinated hydrocarbons (for example, with carbon tetrachloride. chlorobenzene, chloroform and trichloroethylene). Also swollen by esters (for example DBP and DOP), ethers (for example, diethyl ether), asphalt, camphor oil and various aqueous oxidizing agents (for example, dilute nitric acid and potassium permanganate).

10 Not Resistant To

Outside exposure, unless protected by, for example, carbon black or U.V. stabilized. Dissolved by aromatic and chlorinated hydrocarbons at elevated temperatures, for example, 85°C (185°F). Degraded by strong oxidizing agents, for example, oleum and fuming nitric acid (especially when warm). High temperatures, and contact with copper or cuprous alloys, will cause rapid decomposition; however, grades are available which contain certain types of thermal stabilizers to effectively reduce this degradation problem. Not so resistant to oxidation at high temperatures as PE.

11. Material Detection or Identification

With a density of $0.9 gcm^{-3}$ (0.52oz/cu.in) the material (solid, non-filled material) will float in both water and in saturated, magnesium chloride solution. (The addition of 30% glass fiber can raise the density to approximately $1.1 gcm^{-3}$/0.64oz/cu.in). Will float in isophorene but sink in an 80:20 mixture, by volume, of alcohol and water. The natural color of the material is a translucent, ivory white. Components usually feel hard and have a pleasant, dry handle (or feel): may feel slightly waxy. Molded surfaces cannot be readily scratched or scuffed and are virtually unbreakable unless cut.

When this material is heated in a flame then, it softens to give a melt at approximately, 170°C (338°F) and burns readily with a faintly luminous yellow tipped blue flame. There is very little residue and flaring, burning drips are produced together with a candle-like smell. When heated in the absence of a flame then the material will melt at approximately 170°C (338°F) and decomposition will commence at about 300°C (572°F) with the evolution of low molecular weight hydrocarbons. These can be ignited by a flame or radiant heat source and once started the material will continue to burn even if the ignition source is removed; burning is accompanied by flaming, molten droplets. Can be distinguished from other polyolefins by its hardness, density and high melting point. LDPE has a melting point of 110 to 125°C (230 to 257°F), LLDPE 115 to 128°C (239 to 262°F), HDPE is 130 to 135°C (266 to 275°F) and PP is 165 to 175°C (329 to 347°F). Pure isotactic PP although not commercially made as a molding material (has a melting point greater that 170°C (338°F) and a density of approximately $0.91 g.cm^{-3}$ (0.53 oz/cu.in)) is the major ingredient of the molding powders with atactic PP being the minor ingredient. Isotactic PP is insoluble in heptane, or other suitable hydrocarbons e.g. xylene, whereas, atactic PP is soluble. The percentage of material insoluble in heptane is known as the isotactic index and, together with MFR, is used to grade PP materials.

12 Coloring

As the natural color of the material is a translucent, ivory white, then a wide color range is possible. Being a semi-crystalline, thermoplastics material, transparent moldings are not usually possible by injection molding. Best results given by the use of fully compounded material as dry colors and masterbatches can sometimes give rise to streaking, due to dispersion problems.

When using various coloring systems it is often recommended to try to match the particle size or granular shape of the polymer to that of the coloring system. For example, use powder color concentrates or dry colors with powder PP and, masterbatches, of similar particle size to the plastic granules, with granular feedstocks. This is not always the case with dry colorants as the use of powdered PP with a dry colorant necessitate a very long mixing time (i.e. up to 1 hour) to adequately disperse the dry colorant within the mix.

Certain types of blue and green pigments (e.g. phthalocyanine type) can increase the degree of crystallinity in a polypropylene molding. This effect increases the stiffness properties of PP however, it also reduces the impact strength and can cause the molding to distort on cooling as a result of differential shrinkage occurring.

When PP is used for outdoor applications, it requires some form of U.V. stabilization. A well dispersed carbon black at 2.5% concentration is particularly good for both functions i.e. coloring and weather protection. In general, opaque colors can also offer some protection to PP from the effects of weathering, however, they are not as resilient as black. Nickel-based U.V. stabilizers should not be used with pigments that contain traces of sulphur, within the formulation, as discoloration of the color can occur. The most common group of pigments which react in this way is the cadmium based yellow, orange and red types. Use of the Hindered Amine-type light stabilizers (HALS) prevents such discoloration problems occurring.

Various types of masterbatches are used with PP - such as PE based, PP based and the universal type. Addition levels of, and greater than, 3% of the PE-based type can lead to poor weld line strengths and delamination occurring due to poor dispersion, or inhomogeneity, of the PE in the PP molding; typical concentration levels of 1 to 2% are recommended. Universal masterbatch concentration levels of 1 to 2% are also recommended otherwise problems similar to the PE based type occur. When using a high MFR grade of PP to ensure that the correct denseness of color is obtained specially formulated PP masterbatches made be required. These are usually produced from a base PP polymer having the same MFR as the molding material. Liquid coloring is often used giving good results. However care should be taken to ensure that the minimum quantity of color concentrates (i.e. liquid colorant) is used so as to prevent screw slippage, causing erratic screw recovery times to occur.

13 Materials and Components Handling

Available in both powder and granular forms which are usually stabilized to protect the material against decomposition during processing. Will absorb approximately 0.02% water in 24 hours at room temperature. This means that drying is not normally necessary. However, if the material has absorbed a small amount of moisture as a result of long storage it is advisable to dry in a hot air oven for 2 to 3 hours at 80°C (176°F), in a desiccant dryer for 1 to 1½ hours at 90°C (194°F).

This material may not be joined to itself using solvents as there is no solvent at room temperature. Because of its inert, 'non-stick' surface it is difficult to bond using adhesives; some success with contact or hot melt adhesives. If the surface is made polar, for example by using a flame or an electrical discharge, then this material may be bonded to metals using epoxies or nitrile-phenolic adhesives.

14 Mold and Gate Considerations

A suggested mold temperature is 50°C (122°F). If a melt temperature of 240°C (464°F) is used the maximum amount of heat that needs to be removed for each gram of shot is 557 Joules(J). Over the temperature range quoted, the specific heat is approximately 2930 J/kgK.

Altering the mold temperature can alter the degree of crystallinity, shown by the density, and this has a large effect on properties; gives a harder, more impermeable material. Mold temperatures usually lie within the range 40 to 80°C (104 to 176°F); however, chilled molds are sometimes used for the production of thin wall packaging moldings. The value of temperature differentials, in aiding the production of distortion-free moldings, should be remembered. For example, the bowing-in of side walls on box-shaped moldings, may be reduced by running the core of the mold (side walls only) at a lower temperature than the cavity, for example, 25°C (45°F) lower. Can correct some warping by this technique (for example, if the end walls of a molding are tending to bow towards the injection side of a mold) then by running the ejection, or moving half of mold, at a higher temperature this will counteract this.

Typical runner sizes for cold runner molds range from 4mm to 7mm (0.157 to 0.276in). Many different types of gate are used, for example, pin, submarine and edge types. PP does not suffer, as like other materials, from the problem of cracks occurring within the gate area and therefore multiplate molds {using the pin type gate, and self-shearing types (e.g submarine and winkle gates)} are regularly used. Pin gates diameters of as low as 0.7mm (0.028in) can be used, however diameters of 1 to 1.5mm (0.039 to 0.060in) are more typical. Keep the land length as small as is practicable, approximately 0.7mm (0.028in). With edge gating, the minimum gate depth should be half the wall thickness; the width should be at least double the thickness and progressively increase along with the melt flow length in the cavity. The width:thickness ratio of the gate should be 1/25 of the longest flow path length (expressed in mm).

Pinpoint gates are useful with some components, for example those with thin walls, but if high injection speeds are used, be careful that jetting does not occur. Gate into the thickest part of the molding and keep the wall thickness as uniform as possible so as to minimize distortion. For large flat moldings, where distortion may be a problem, use multiple gates laid out so as to give uniform mold filling (high melt and mold temperatures minimize distortion).

Good venting is also essential, for example, using a ring vent which surrounds the cavity and which is 0.025 to 0.038mm (0.001 to 0.0015in) deep and with a land length of 1.5mm (0.060in).

To avoid sink marks, use large, full round sprues (also full round runners) and keep the size of the ribs small, for example, 50 to 60% of the adjacent wall thickness (t) (ie (0.5 to 0.6t)). Distortion of large, flat areas, for example box walls, may be countered by deliberately increasing the wall thickness, in the center, by incorporating an external rib. If a molded-in hinge is incorporated, keep the hinge-line straight and generously radius the wall thicknesses where the reduced section, which forms the hinge, is formed. The thickness of the hinge is usually determined by the thickness needed to obtain reasonable flow across the hinge section. Often lies within the range 0.25 to 0.6mm (0.010 to 0.024 in). Locate all gates on one side of the hinge, in a position which will ensure that full pressure is behind the melt as soon as it begins to flow through the hinge restriction.

If the ejectors are made too small, then white stress marks

will be produced so, position ejectors so that they operate on ribs, bosses or under side walls; make them as large as possible and give as large a taper as possible. However, draft angles can be low if required, for example, 0.5 degrees; when using narrow draft angles always ensure that the maximum area of molding is being used to eject the component otherwise distortion may occur.

PP is widely used for caps and closures etc. because undercuts can be easily 'bumped', or jumped off during component ejection. When ejecting a component by this method, it is essential to allow sufficient space so that the component can expand either in an inwards or outwards direction. This can be achieved by various methods, the most common being the two stage ejection technique. A typical depth of undercut that can be readily 'jumped' is 0.5mm (0.02in). Glass filled PP can also be used for undercuts but due to it's increased rigidity the depth of undercut that can be used is only 0.25mm (0.01in).

Components possessing wall sections of 3mm (0.118in) and above, when produced from a homopolymer suffer from cavitations, (the formation of bubbles, within the section). This is due to high shrinkage and to, variation in melt cooling rates across and along the section thickness. To consistently produce bubble free moldings, by normal processing methods, is extremely difficult and attempting to overcome such problems often results in high reject levels. If this is the case, then it is preferable that the homopolymer is replaced with a PP copolymer which contains a small percentage of ethylene or a specific CR grade in order to overcome this problem.

As PP is notch sensitive, the type of textured surface used for molded components should be carefully chosen. Although a textured surface readily disguises surface blemishes, or markings, it is prone to reducing the impact properties of the molding (i.e. depending upon the depth and style of texture). The presence of the textured surface can act as a notch and therefore, if the component is struck on the side opposite the notch (i.e. on the non-textured side) then failure can occur.

Particular care must be taken when using molded-in brass inserts, as the degradation of PP can readily take place due to traces of copper present in the alloy - used to form the brass. If brass inserts are to be used, then it is preferable to plate them with materials similar to those used for steel inserts e.g. nickel. Most inserts need to be degreased and preheated (up to 120°C (248°F)) prior to being placed in the mold.

PP is readily use for runnerless molds due to its wide processing range and good thermal stability. The comments on runnerless molds for this material, are very similar to that given for HDPE.

15 Flow Path: Wall Thickness Ratio

This material is an easy flowing material; at 1mm (0.039in) wall thickness the flow path: wall thickness ratio can reach up to 340:1 for a high MFR grade (typical grade 230.1).

16 Projected Area

Usually 1.0 to 2.5tsi. (15 to 39MNm^{-2}) is sufficient. For components of thin wall sections and requiring a long flow length this may have to be increased to 3.5tsi or 54MNm^{-2}.

17 Cylinder Equipment

General purpose screws can be used, for example, with an L:D ratio of about 20:1 and a constant pitch of 1D. The feed zone should be approximately 0.5L, the compression zone 0.3L and the metering zone 0.2L; a compression ratio of about 2:1 should be sufficient. The screw is usually fitted with a non-return valve as it's use is essential when long screw forward times are needed so as to obtain sink free moldings on thick sectioned components. Valves are not essential on the nozzle unless high back pressures are involved (may be necessary to ensure that good pigment dispersion occurs when using an on-line coloring technique). The best results have been obtained by spring loaded or hydraulically operated types; decompression, or suck-back, is also used. Foamable PP molding compounds can be used on conventional screw type machines; the use of a shut off nozzle is necessary to prevent the foamed material from leaving the cylinder prematurely. Since PP can be adversely affected by contact with copper or copper alloys, extension nozzles made from such materials, should be chromium or nickel plated before use. Most types of extended and hot runner nozzle work well provided that there is no excessive restriction to the melt flow. As PP sets at high temperatures, ensure that the heat losses are kept low and that the temperature is accurately controlled.

Component weight variation is often overcome by using a dedicated PP designed screw. Such a screw can give more consistent screw recovery times, and better melt homogeneity, particularly for the manufacture of thin walled components which necessitate long melt flow lengths. Changes to the depth of flights at the feed section and screw pitch often overcome the problems associated with the GP design of screw. High hopper throat temperatures e.g. 80°C (176°F) can also contribute to a more consistent plasticizing and screw recovery when using a standard GP screw. The use of short nosed torpedoes (ie having an inclusive angle of 60 degrees) of the back flow valve assembly are often preferred to the long nosed type to overcome melt cushion and component weight variation.

18 Screw Cushion

On smaller machines this is approximately 3mm (0.118in); on larger machines 9mm (0.354in) is suggested.

19 Shot Capacity

Because of the material's good heat stability as low as 20% of the cylinder's rated capacity may be used: the maximum possible is 85%. Due to the reduction in melt density (ie $0.551 g/cm^3$)) as compared to PS (0.95 $g.cm^{-3}$/0.55 oz/cu.in). The cylinder capacity has to be down rated by 22% e.g. if the shot capacity for a given screw and barrel assembly is given as 160g in PS, then the actual shot capacity for PP will be 126g.

20 Melt Temperature

This is as measured in the nozzle or by an air shot technique; 220°C to 275°C (428 to 527°F). Do not exceed 275°C (527°F) if possible as the viscosity changes rapidly and oxidation may occur.

21 Barrel Residence Time

More susceptible to oxidation than PE as PP contains many tertiary carbon atoms. At a temperature of 260°C (500°F) the maximum residence time in the barrel should not exceed 5 to 6 minutes. When the temperature is 270°C (527°F) then the maximum residence time in the barrel should not exceed 2 to 3 minutes.

22 Temperature Settings

Zone No.	Location	Temperature °C FROM	TO	Temperature °F FROM	TO
0	Hopper throat	30	80	86	176
1	Barrel rear	190	230	374	446
2	Barrel middle	200	240	392	464
3	Barrel middle	210	260	410	500
4	Barrel front	220	270	428	518
5	Nozzle	220	270	428	518
6	Mold	30	80	86	176

Please note that it is the melt temperature which is important;

those temperatures given in the table are only suggested, initial settings. The temperature of the hydraulic oil, the hopper block and the material temperature should also not vary excessively. Mold temperature control is extremely important and therefore these temperatures must be checked regularly.

23 Injection Speed

To minimize internal stresses and distortion, fast injection speeds are used, however such speeds may not be possible with some grades (because of cavitation problems) or, with some molds. A high speed, with a textured finish, can give alternate glossy and dull areas which are often in the form of 'fingers' radiating out from the feed points. Use a slow speed and higher mold temperatures if this occurs. Machines having the facility to profile the injection speed are a considerable asset when molding PP.

24 Injection Pressure

The machine should be capable of giving up to:

first stage, up to 1800bar; 180MNm^{-2}; 26,100psi.
second stage, up to 1500bar; 150MNm^{-2}; 21,700psi.

High dwell pressures may be needed to avoid cavitation, or excessive shrinkage, but be careful to avoid over-packing.

25 Screw Rotational Speed

Adjust speed (the rpm) to suit the molding cycle but do not use a screw surface speed of greater than 0.75 to 0.8ms^{-1} (2.46 to 2.62 ft/sec). If a screw of 40mm (1.57in) diameter is being used then the rpm should not exceed 380. Similarly the maximum diameter of screw when rotated at 200rpm, in order to obtain a screw surface speed of 0.75 to 0.8ms^{-1} (2.46 to 2.62 ft/sec) is 76mm (2.99in).

26 Back Pressure

Up to 200bar; 20MNm^{-2}; 2,900psi should be available. Should be kept as low as possible: usually a nominal value of 5 bar (0.5MNm^{-2}, 72.5psi) is used so as to achieve consistent shot weights. May need to be increased, to assist dispersion, when using colorant but keep as low as possible.

27 Shutting Down

No need to purge with another material when closing down.

Usual procedure is to purge barrel clean, leave the screw in its forward position and switch off the heats.

28 Reprocessing

When this material is reclaimed it is suggested that up to 15% be blended into the virgin material. If the material has been recycled too many times then it will degrade. This is shown by a increase in MFI, an increase in brittleness, excessive odor from the moldings and signs of decomposition, for example, burn marks or brown streaks.

29 Finishing

The surface of this material may be made more receptive to inks, paints, lacquers or to adhesives by various forms of surface pretreatment such as a corona discharge, flame and chemical treatment. Various types of chemicals are used to etch the surface of polypropylene moldings; these range from potassium permanganate, chromic acid and trichloroethylene vapors. Each chemical is carefully prepared and the moldings are dipped in the prepared solutions at a temperature of between 60 to 70°C (140 to 158°F) for a period of 30 seconds to 15 minutes. The type of chemical used and duration of dip is associated with the subsequent decorative process to be used on the molding. Specially formulated primers can be used to form a adhesive coating to the molded surface, however this technique is not as effective as the chemical pretreatment process. Components can be decorated by printing, painting, hot foiling or vacuum metallizing. Electroplated PP moldings exhibit better thermal cycling properties than ABS moldings. In-mold foiling can also be used to decorate moldings made from PP; particularly used in the make-up and cosmetic field.

Because of PP's excellent resistance to solvents the use of solvent based adhesives is limited and therefore other forms of jointing methods such as snap fits, the use of self tapping screws and welding methods (such as hot plate or shoe, friction and ultrasonic) are often preferred. When hot plate welding PP, it is usual to coat or cover the hot plates with PTFE so as to prevent the material sticking to the surfaces of the hot plate.

PP can be readily machined with cutting tools used for metals; use high cutting speeds and low feed rates. The edges of the cutting tools must be sharp in order to achieve a good surface finish; use plenty of cutting fluid to cool the cutting surface and prevent any heat build up when machining. Typical operations that can be readily performed are milling, drilling and tapping, turning sawing and planing. PP cannot be readily surface ground and polished; specialized techniques are necessary to produce machined surfaces to a very high accuracy and quality.

30 Other Comments

Foamed or expanded PP is being increasing used for many applications due to it's lightness and good structural properties. The Cimpres process, is, for example, used with PP where thick sectioned moldings can be produced without the usual problem of sinks and long cycle times. A great deal of PP is used to manufacture crates and with such products creep can be a problem due to the moldings being overstacked; the product may become distorted or the stack may collapse.

31 Typical Components

With a combination of light weight, toughness, high temperature resistance, rigidity and excellent resistance to chemical attack, PP is suitable for a wide range of components produced by injection molding. These include automobile fascias, bumpers, bottle crates, washing machine tubs, kitchen ware, textile bobbins, tool handles and domestic waste systems. However, despite this materials high stiffness, it is still subject to creep and it should not be used in applications where it will be continuously loaded under high stress.

By taking advantage of the integral, hinge effect, components such as boxes, and their lids. can be produced in one shot. Molded hinges often show a white line along their center as a result of orientation; it is not a weakness and is more pronounced for PP-CO. PP-HO gives slightly superior hinges. Use easy flow grades, and do not use filled grades, for this application.

PP is now widely used in the automotive industry, particularly in Japan, where it is extensively used for bumper fascias and instrument panels. Often such panels are based on filled compounds, for example, on talc filled or, on talc/glass fiber filled compositions. PP with high levels of crystallinity are also becoming more popular as this high level of crystallinity gives thermoplastics materials with improved surface hardness, better scratch resistance and higher temperature resistance. Such materials are used above the belt line, in automotive interiors, for pillar and quarter panel trim. Filled grades, based on new PP materials, are being used in other industries as, for example, washing machine tubs, washing machine bodies refrigerator doors and dishwasher doors. Such applications take advantage of the good, hot water resistance of PP: special grades are available for such applications.

32 Unit Price

If the price of a natural, unmodified grade of GPPS is taken as one then the relative price of this molding material is approximately:

Natural Homopolymer PP GP grade	0.81
Very easy flow grade	0.83
Natural Copolymer PP GP grade	0.87

The price relative to GPPS can also be found in Table 7.

RUBBER MODIFIED POLYPROPYLENE

1 Common name(s)

Rubber reinforced polypropylene: polyolefin-rubber blends.

2 Abbreviation(s)

RRPP : RMPP : PP/EP(D)M : EP(D)M-PP : EMT : EMPP : IMPP : OTE : TPO.

If the rubber content of a PP blend is less than 50%, then the material may be referred to as rubber reinforced polypropylene (RRPP) or, as rubber modified polypropylene (RMPP) or, as an elastomer modified thermoplastic (EMT) or, as an olefin thermoplastic elastomer (OTE) or, as impact modified polypropylene (IMPP). The material may also be known as a thermoplastic elastomer (TPE) or, as a thermoplastic rubber (TPR). The rubber/elastomer in such a mixture is not crosslinked, or cured, or vulcanized). The term rubber reinforced PP (RMPP) will be used for those systems in which the rubber is present in relatively low concentrations (<30%) and is not cross-linked/cured.

3 Alternative names

Rubber modified polypropylene ; polypropylene/ethylene propylene rubber blends; ethylene propylene rubber, polypropylene blends; elastomer modified thermoplastic; elastomer modified polypropylene; impact modified polypropylene; olefin thermoplastic elastomer; thermoplastic polyolefin rubber.

4 Some suppliers 5 Trade names or trade marks

(a) Polypropylene (PP)

Supplier	Trade name
Advanced Global Polymers	Adpro
Amoco	Amoco PP
Apryl	Apryl
Atochem	Lacqtene P
Bamberger Polymers	Bergaprop
B.A.S.F.	Novolen.
B.P	Napryl
Enron/Delong Prochem	Norchem
Huls	Vestolen P
Cabot Plastics	Cabelec (conductive PP)
Chemie Linz	Daplen PP
DSM	Stamylan P
El Paso	Rexene

Himont	Moplen, Profax
Hoechst/Hoechst Celanese	Hostalen PP
I.C.I.	Propathene
Laporte Industries	Eltex P
Monmouth Plastics	Empee PP
Neste	Neste PP
Norchem Inc	Nortuff
Petrofina	Finaprop
Phillips Petroleum Ltd.	Marlex
Schulman	Polyflan
Shell	Shell PP
Soltex	Fortilene
Solvay	Eltex P
Statoil	Statoil PP

(b) Ethylene propylene rubber/elastomer {EP(D)M}

B.F. Goodrich	Epcar
Bunawerke Huls	Buna AP
D.S.M.	Keltan
D.S.M.	Idemitsu
DuPont	Nordel
Enimont	Dutral and Intolan
Esso	Vistalon
Exxon	Vistalon
Mitsui Petrochemicals	EPT
Polysar	Polysar EPDM
Sumitomo	Esprene
Total	EP Total
Uniroyal	Royalene

(c) Compounded material

Colonial	Somel
D.S.M.	Kelprox
D.S.M.	Keltan TP
Dynamit Nobel	Dynaform
Enimont	Moplen
Esso	Vistaflex PP
Exxon	Trefsin
Ferro Corp	Ferroflex
Himont	Profax
Hoechst	Hostalen
Huls	Vestolen EM
I.C.I.	Propathene OTE
Republic Plastics	Eta
Research Polymers	Ren-flex
TelcarSchulman	Polytrope
Uniroyal	Uniroyal TPR

(d) Dynamically vulcanised material

| Monsanto | Santoprene |
| Monsanto | Levaflex |

6 Material Properties

This class of TPEs consists of blends of the thermoplastic PP with an unvulcanized ethylene propylene rubber/elastomer. May contain smaller amounts of other materials, for example, HDPE, plasticizers and carbon black.

Blends of rubbers and plastics, which give rubber-like materials, can be roughly divided into two categories. These are, those in which the plastics material has a low level of crystallinity and which derive their rigidity from the base plastics material: this has a relatively high glass transition temperatures (Tg), for example, PVC. The second class is based on thermoplastics materials which have a high level of crystallinity, for example, polypropylene (PP) or its copolymers.

PVC, and its copolymers, are most often converted to TPE-type materials by melt mixing them with plasticizers (plasticisers) which are compatible with the plastics material. Both polymeric and low molecular weight plasticizers are used with PVC to give elastomeric compounds. With the semi-crystalline, thermoplastics materials, it is necessary to mix them with rubbers/elastomers, with which they are not molecularly compatible, so as to form the required softer, elastomeric materials. This is because plasticizers are not compatible with a semi-crystalline, thermoplastics material.

When ethylene is copolymerized with propylene, the amorphous, rubbery material called ethylene propylene rubber (EPM – from ethylene propylene monomers) results if the ethylene content is high: commercial polymers may contain 30 to 50% of ethylene units. In order to obtain a sulfur vulcanisable material, a small percentage of a third monomer (3 to 10% of a diene monomer, for example, ethylidene norbornene) is incorporated so as to make a terpolymer; this terpolymer may be referred to as EPDM. An abbreviation which covers both types of ethylene propylene rubber is EP(D)M; usually, all types of this rubber have a propylene content of between 25 and 55% by weight.

Both types of ethylene propylene rubber (also known as EPR), are used to make blends, with polypropylene, as they are both relatively cheap, have good weathering and chemicaal resistance, excellent dielectric characteristics and give, PP-based materials with an acceptable balance of properties. These include, for example, good low temperature impact strength, adequate rigidity at elevated temperatures and, perhaps most importantly of all, the ease of processing associated with thermoplastics materials. That is, the blends have the ability to be processed on thermoplastics equipment - just like any other thermoplastic.

RRPP compounds, in which the rubber is not cured, are best produced by intensively melt mixing the plastic and the rubber (up to 40%) together, for example in an internal mixer, at a temperature greater than the melting point of the PP. A two phase structure should result, in which the rubber is in the form of 0.5 to 5 micron particles which are embedded in a continuous phase of PP. Highly isotactic, polypropylene (PP-H or PP-HO) may be used or, block copolymers of propylene and ethylene (PP-B) (such copolymers are also known as PP-K or, as PP-C or, as PP-CO). The best balance between, for example, stiffness (E modulus), cold temperature impact strength and ease of flow, is obtained by the use of high molecular weight, amorphous EP(D)M and PP-B of medium viscosity. The use of a copolymer, in place of a homopolymer, can reduce the rubber content by approximately 10%.

To improve the heat distortion temperature and stiffness, fillers may be used: for example, by using approximately 25% talc (of particle size 3 to 15 microns) a harder, stiffer compound results. Talc filling is detrimental to cold impact behaviour and to antioxidant performance. By ringing the changes on the types, and quantities, of materials used, it is possible to produce a very wide range of compounds or blends.

EPR rubbers/elastomers are not the only rubbers used. For example, using nitrile rubber in place of an ethylene propylene rubber gives a more oil resistant material (Geolast). Natural rubber is also used (Vyram).

Considerable improvements in properties result, compared to RRPP, if cross-linking/curing is introduced into the rubber particles. PP modified, rubber compounds. Such materials may be known as dynamically crosslinked polyolefin elastomer or TPO-XL or, as a dynamically vulcanized material or, as an elastomeric alloy (EA) or, as an elastomeric alloy thermoplastic vulcanizate (EA-TPV). If enough plastic is present (say 50%), the compound can still be easily processed by thermoplastic techniques.

Compared to compounds which contain non-crosslinked rubber, the dynamically vulcanized (DV) materials have substantially improved compression set, permanent set resiliency and high temperature shape retention. They are a very rapidly growing sector of the thermoplastic elastomer industry.

7 Ease of Flow

MOLDFLOW DATA FOR EXXON VISTAFLEX

Grade.	Viscosity (Nsm^{-2}) at $1,000s^{-1}$ at the following temperatures; in °C (°F)		
	220°C (428°F)	240°C (464°F)	260°C (500°F)
General Purpose grade (240G)	146	129	119
Easy flow grade (206B)	109	94	84

Material Grade	Viscosity (Nsm^{-2}) at 220°C (428°F)			
	$100s^{-1}$	$1,000s^{-1}$	$10,000s^{-1}$	$100,000s^{-1}$
Vistaflex (240G)	623	129	26	8
Vistaflex (206B)	352	94	21	7

There is a big difference in viscosity between the rubber and the plastic and this can lead to mixing problems (even the lowest molecular weight EP(D)M has a higher viscosity than any grade of PP. However, even the addition of 30% PP, greatly decreases the viscosity of the rubber. If only a rubber addition is being considered, for example, to reduce cavittation during injection molding, then it may be best to use high rubber content masterbatches – particularly if mixing is done on the injection molding machine. At low rubber addition levels, the blends process like PP.

8 Shrinkage

The addition of rubber to PP reduces the shrinkage by approximately 10% and reduces voiding (cavitation) in thick sections. Shrinkage is of the order of 0.018in/in or mm/mm ie 1.8% for RRPP and for the DV type. The addition of fillers, for example talc, reduces the shrinkage to approx. 1.2% at 40% loading for RRPP.

9 Resistant to the Following

RRPP is resistant to oils, fats, alcohols and glycols; also to caustic alkali, strong and weak acids, and low to medium strength oxidising chemicals. Also resistant to most chemicals of a polar nature, for example, water. In general, EP rubbers will not swell when in contact with highly polar liquids (such as water) but swelling may occur with aliphatic, non-polar or slightly polar liquids. EP rubbers have excellent resistance to ozone, oxygen, water, dilute acids and alkalis.

The major reason for adding rubber to PP plastics is to overcome the poor low temperature properties; some blends are impact resistant down to sub-zero temperatures, for example, –40°C (–40°F). One reason for this is that the rubber stops large spherulite growth; large spherulites are stress concentration sites in PP.

10 Not Resistant to

When the rubber is unvulcanised, the addition of rubber worsens the chemical resistance of the PP; chemical resistance worsens as rubber content increases. RRPP mouldings have relatively poor resistance to petrol and other hydrocarbon solvents. The resistance of EP rubbers to aliphatic and aromatic solvents is poor: also has poor resistance to animal and vegetable oils but good resistance to hydraulic fluids.

11 Material Detection or Identification

With an SG of less than one, the material (solid, non-filled material) will float in both water and in saturated, magnesium chloride. At small rubber addition levels, the blend will have a similar SG to PP as the SG of ethylene propylene rubbers is approximately 0.86. Many RMPP copounds contain 10 to 30% of EPDM and so their densities will be less than that of PP.

The hardness range of this class of materials is from 60 Shore A to 60 Shore D. Hardness, modulus and tensile strength all increase as the PP content increases. (The hardness of rubber compounds is typically measured by the Shore Durometer test: an indentor is pressed into the material, under a specified load, and the amount of penetration determines the hardness. A circular tip is used for the A scale and a pointed tip is used for the D scale. D means harder materials.)

Transparent mouldings are not possible for mixtures of PP and EPDM. However, the natural colour of the materials is similar to that of PP (i.e. an off-white), and so a large colour range is possible. However, it is not often seen as these materials are often used in automotive applications where good outdoor stability is essential, This may be achieved relatively cheaply by the addition of carbon black (a suitable grade is a furnace black of 20 to 25 millimicrons at a loading of 1 to 2%).

The mixed material feels softer and more flexible than PP and if the rubber and the plastic are not well mixed together then, on bending or flexing, the mouldings will easily break. Unless protected, very easy to scratch with a knife. When this material is heated in a flame then, it burns easily without any real distinctive odor; burning drops are often formed. As the rubber content increases, then the hardness, modulus (stiffness), and Vicat softening point decrease. The DV materials are usually softer and have a pronounced, rubbery feel: the softer grades feel like a piece of vulcanized rubber.

The amount of unvulcanized rubber, in a blend, may be approximately determined by extraction with hot xylene; after extraction, cool, and then determine the soluble matter in the cold xylene. A density measurement will show if there is filler present and an ash content will show the amount.

12 Coloring

The natural colour of the material is similar to that of PP (i.e. an off-white), and so a large colour range is possible. However, this large color range is not often seen as these materials are often used in automotive applications where good outdoor stability is essential, As this may be achieved relatively cheaply by the addition of carbon black then, many PP/EPDM materials are black.

May, for example, be supplied as natural grades or as black grades. Such materials may howver, be usually coloured using pigments or colourants suitable for polyolefins, for example, by using solid masterbatches. In the case of U.V. stabilized grades, attention should be paid to some organic colourants which adversely affect the long-term aging action of some stabilizers, for example, the hindered amine types.

13 Materials and Components Handling

If the rubber and the PP have been compounded previously together, then this melt compounded material, in pellet form, may be treated as though it were PP. Many RRPP materials contain 10 to 30% of unvulcanized rubber and such materials do not usually absorb significant amounts of water. This means that drying is not normally necessary. However, if the material has absorbed a small amount of moisture as a result of long storage it is advisable to dry in a hot air oven for 2 to 3 hours at 80°C (176°F), or in a desiccant dryer for 1 to $1^{1}/_{2}$ hours at 90°C (194°F).

If the individual materials (i.e rubber and PP) are being fed to the injection molding machine (sometimes known as 'direct blend injection molding' or, DBIM) then ensure that component separation does not occur during feeding. This may be avoided by metering the PP and the rubber directly into the throat of the injection molding machine. To have any chance of success, good mixing within the injection molding machine is essential. This may be achieved, for example, by using a mixing screw and/or a mixing head.

Machining of moldings made from this plastic/rubber blend, should take account of the soft, resilient nature of this material by, for example, supporting the molding well, using sharp cutting tools and using adequate cooling. Not usually joined with adhesives; may be heat welded.

14 Mold and Gate Considerations.

A suggested mold temperature is 50°C (122°F). If a melt temperature of 240°C (464°F) is used then the maximum amount of heat that needs to be removed for each gram of shot is 557 Joules(J). Over the temperature range quoted, the specific heat is approximately 2264J/kgK.

It is suggested that the feed system be sized and, laid out, so as to minimise pressure losses; this is because the shrinkage of this material is still relatively high and so dwell pressure will need to be applied for a long time. Be careful, however, not to overpack.

Typical runner sizes for cold runner molds range from 4mm to 7mm (0.157 to 0.276in). Many different types of gate are used, for example, pin, submarine and edge types. PP-EPDM blends do not suffer, as like other materials, from the problem of cracks occurring within the gate area and therefore multiplate molds {using the pin type gate, and self-shearing types (e.g submarine and winkle gates)} are regularly used. Pin gates diameters of as low as 0.8mm (0.032in) can be used, however diameters of 1 to 1.5mm (0.040 to 0.060in) are more typical. Keep the land length as small as is practicable, approximately 0.7mm (0.028in). With edge gating, the minimum gate depth should be $0.7t$ where t is the wall thickness; the width should be at least double the thickness and progressively increase along with the melt flow length in the cavity.

Pinpoint gates are useful with some components, for example those with thin walls, but if high injection speeds are used, be careful that jetting or surface delamination does not occur. Gate into the thickest part of the molding and keep the wall thickness as uniform as possible so as to minimize distortion. For large flat moldings, where distortion may be a problem, use multiple gates positioned so as to give uniform mold filling and equal pressure drops (high melt and mold temperatures minimize distortion).

As mold filling speeds can be high good venting is also essential, for example, using a ring vent which surrounds the cavity and which is 0.025 to 0.038mm (0.001 to 0.0015in) deep and with a land length of 1.5mm (0.060in).

Removal of the feed system may be accomplished very easily, for example, with a sharp knife or scalpel.

To avoid sink marks, use large, full round sprues (also full round runners) and keep the size of the ribs small, for example, 50 to 60% of the adjacent wall thickness. Distortion of large, flat areas, for example box walls, may be countered by deliberately increasing the wall thickness, in the center, by incorporating an external rib. If a molded-in hinge is incorporated, keep the hinge-line straight and generously radius the wall thicknesses where the reduced section, which forms the hinge, is formed. The thickness of the hinge is usually determined by the thickness needed to obtain reasonable flow across the hinge section. Often lies within the range 0.25 to 0.6mm. Locate all gates on one side of the hinge, in a position which will ensure that full pressure is behind the melt as soon as it begins to flow through the hinge restriction.

These materials are softer than PP and so greater care should be taken over ejecting the moldings ,for example, a large number of big area ejectors should be used on the very soft grades.

If the ejectors are made too small, then white stress marks will be produced so, position ejectors so that they operate on ribs, bosses or under side walls; using as large a taper as possible. However, draft angles can be low if required ,for example, 1.0 to 1.2 degrees (incl.); when using narrow draft angles always ensure that the maximum area of molding is being used to eject the component otherwise distortion may occur. Larger draft angles are (5 to 10°) are necessary for mouldings with textured surfaces.

PP-EPDM is used for mouldings that require the undercut to be easily 'bumped', or jumped off during component ejection. However, this does depend upon the amount of rubber content used. When ejecting a component by this method, it is essential to allow sufficient space so that the component can expand either in an inwards or outwards direction. This can be achieved by various methods, the most common being the two stage ejection technique. A typical depth of undercut that can be readily 'jumped' is 0.6mm (0.024in).

Components possessing wall sections of 3mm (0.118in) and above, can be readily molded using PP-EPDM blends. Cavitation is not such a problem with PP-EPDM blends as encountered with the homopolymer PP. PP-EPDM blends are also not as notch sensitive as PP, therefore can be used for moldings requiring a range of surface textures. As a consequence of this property these materials have replaced PP for many applications.

Particular care must be taken when using molded-in brass inserts, as the degradation of the PP content within the blend can readily take place due to traces of copper present in the alloy - used to form the brass. If brass inserts are to be used, then it is preferable to plate them with materials similar to those used for steel inserts e.g. nickel. Most inserts need to be degreased and preheated (up to 120°C (248°F)) prior to being placed in the mold.

PP-EPDM is readily use for runnerless molds due to its wide processing range and good thermal stability.

16 Projected Area

Similar to that used for PP, for example, 2 to 5tsi., 41 to 70MNm^{-2}, is usually sufficient. The actual value is dependent upon the amount of rubber content used.

17 Cylinder Equipment

Nozzles are usually of the open type and the screw usually has a back-flow valve. Decompression may prove advantageous when molding easy-flow materials (high MFR or MFI). Most types of extended and hot runner nozzle work well provided that there is no excessive restriction to the melt flow. As the base PP, sets at high temperatures, ensure that the heat losses are kept low and that the temperature is carefully controlled. A general purpose designed screw is often used to process PP-EPDM compounds giving excellent results; typical L:D ratios 20:1 and a C.R. of 2.2 to 2.3:1 are used. When pellets of PP and EPDM are being fed to the machine then, best results will be obtained if the injection molding machine is operated so as to improve mixing, for example, with high back pressures. it will probably be necessary to use a special screw (one designed for mixing, for example, with mixing pins along its length) or, a vented machine. (Vented machines, because of their screw geometry, function well as mixers).

18 Screw Cushion

About 4mm (0.157in) on small machines and 9mm (0.354in) on large machines.

19 Shot Capacity

Up to approximately 80% of the rated capacity in PS may be used.

20 Melt Temperature

This is as measured in the nozzle or by an air shot technique; for example, 220 to 270°C (428 to 518°F). Higher temperatures can give longer flow lengths and better weld strengths if mulitple gates are being used. The use of low temperatures can however, improve dispersion. As too high a melt temperature can cause decomposition, or degradation, it is best to process at approximately 250°C (482°F).

21 Barrel Residence Time

At a temperature of 280°C (536°F) the maximum residence

time in the barrel should not exceed 5 to 6 minutes. When the temperature is 290°C (554°F) the the maximunm residence time in the barrel should not exceed 1 to 2 minutes.

These materials are thermally stable - even the DV type materials are relatively stable and short production interruptions can be tolerated wtihout purging. If the interruption is longer than 30 minutes then the cylinder, or barrel, should be purged.

22 Temperature Settings

Zone No.	Location	Temperature in °C FROM	TO	Temperature in °F FROM	TO
0	Hopper throat	40	60	104	140
1	Barrel rear	170	210	338	410
2	Barrel middle	210	230	410	446
3	Barrel middle	220	250	428	482
4	Barrel front	230	260	344	500
5	Nozzle	240	270	464	518
6	Mold	20	60	68	140

Please note that it is the melt temperature which is important; those temperatures given in the table are only suggested, initial settings. The temperature of the hydraulic oil, the hopper block and the material temperature should also not vary excessively. Mold temperature control is extremely important and therefore these temperatures must be checked regularly.

23 Injection Speed

High speeds needed but may not be possible with some grades (because of cavitation problems) or, with some molds. A high speed, with a textured finish, can give alternate glossy and dull areas which are often in the form of 'fingers' radiating out from the feed points. This fault is particularly noticed with hot runner molds. Use a slow speed and higher mold temperatures if this occurs. When multiple-gated molds are being used, it is not necessary to use very fast filling speeds in order to get high weld strengths; employ a steady filling speed through large gates and do not overpack.

24 Injection Pressure

The machine should be capable of giving up to:
first stage, up to 1800 bar; $180 MNm^{-2}$; 26,100psi.
second stage, up to 1500 bar; $150 MNm^{-2}$; 21,700psi.

25 Screw Rotational Speed

Should be adjusted to suit the molding cycle but do not use a screw surface speed of greater than 0.55 to 0.65ms^{-1} (1.80 to 2.31ft/sec). If a screw diameter of 40mm (1.57in) is being used then the rpm should not exceed 310. Similarly, the maximum diameter of a screw when rotated at 200rpm in order to obtain a screw surface speed of 0.55 to 0.65ms^{-1} (1.80 to 2.31ft/sec) is 62mm (2.44in).

26 Back Pressure

Should be kept as low as possible: usually a nominal value of 5 bar (0.5MNm^{-2}, 72.5psi) is used so as to achieve consistent shot weights. May need to be increased, to assist dispersion, when using colourant but keep as low as possible. Up to 50 bar; 5MNm^{-2}; 725psi. should be available

27 Shutting Down

No need to purge with another material when closing down. This type of material is quite useful as a 'scouring-type material' used, for example, to clean a barrel of an undesirable color or material.

28 Reprocessing

May be reclaimed on equipment suitable for PP; reclaim is completely reusable with negligible impairment of properties. However, when reprocessing be careful as slight color changes can occur, for example, with RRPP if the material has been overheated.

When this material is reclaimed it is suggested that up to 30% be blended into the virgin material.

29 Finishing

Moldings are painted not only for the decorative effect but also to improve the scratch resistance of the RRPP. Pretreat before painting, for example, by degreasing and priming. Vapor degreasing is accomplished with 1,1,1 trichlorethane; dust may be removed by blowing with ionised air. The surface may then be primed with a system based on chlorinated PE disolved in toluene and then painted with a two component polyurethane systems. Best to avoid the use of silicone mold releases if the injection moldings are to be painted. When molded correctly, parts release easily from the mold. The use of mold release agents is neither required nor recommended in normal operation.

The adhesion of paint inproves as the EPDM content increases. Pretreatment processes include flaming, corona discharge, hot chromic acid, sanding and proprietary primers.

30 Other Comments

Annealing of the molding at 130 to 140°C (266 to 284°F), for 1 to 2 hours, can be an efficient way of improving and stabilizing the physical properties, for example, there is a significant improvement in E-modulus, impact strength and low temperature brittleness of some materials.

31 Typical components

Polyolefin-rubber blends are low cost, two phase systems which show some rubbery behavior. The most widely used blends contain more thermoplastics material (PP) than rubber/elastomer and are basically rubber modified PP.

In recent years the major application of these materials has been in the automotive industry, for example, in applications where the service requirements are not very demanding. Comparative lightness (density 0.87 to 0.9gcm^{-3}(0.50 to 0.52oz/cu.in)), low cost and a wide range of mechanical properties are the major factors responsible for the continuous growth in consumption of RRPP. That is, the materials are versatile as a very large range of grades can be produced by varying the ratios of PP to EPDM, the types of material used, the additives employed etc. Such blends are easy to produce on say twin screw extruders. In general, as the rubber content increases there is an increase in impact strengh, but a decrease in modulus, hardness, dimensional stability on heating and in Vicat softening point (VST).

The main use is in car bumpers, either using flexible mouldings mounted on a steel framework, self supporting mouldings, for example, complete front ends. Car door protection strips are another suitable compounds contain from 10 to 40% rubber- according to flexibility and impact requirements.

Talc containing compounds are used to make car instrument panels and dash boards as they have sufficient stiffness, shape stability, and an impact resistance sufficient to avoid splintering in medium impact crashes. Because these materials fail in a ductile manner, they are safer for the car occupants

In general, mouldings made from melt compounded material, for example made in an internal mixer, exhibit a higher Vicat softening point, a higher elongation at break and better low temperature properties than other mouldings made from simple tumble mixes. To get the best results from any of these compounds, try injection molding them using those conditions suggested for ABS which is to be plated. For example use high melt and molding temperatures, slow filling speeds and the minimum of overpacking.

Major limitations of this class of materials are associated with surface properties and heat resistance. Compounds usually have poor hardness, scratch resistance and limited heat resistance (particularly under load). The addition of peroxides, to give partial crosslinking helps and, surprisingly, can improve the flow.

32 Unit Price

If the price of a natural, unmodified grade of GPPS is taken as one then the relative price of this molding material is approximately:

Automotive grade = 1.36

The relative price to GPPS can also be found in Table 7.

PLASTICIZED POLYVINYL CHLORIDE

1 Common name
Plasticized PVC.

2 Abbreviation(s)
PPVC ; PVC-P .

3 Alternative names
Plasticised polyvinyl chloride : soft PVC. PVC may also be referred to as poly(vinyl chloride) or as poly(monochloroethylene).

4 Some suppliers   5 Trade names or trade marks

Supplier	Trade name
Alpha Chemicals and Plastics	Vynite (PVC/NBR)
Alpha Chemicals and Plastics	Vythene (PVC/PU)
Atochem	Lucalor
Atochem	Lucorex
BASF	Vinidur
BASF	Vinoflex
BF Goodrich	Geon
Blane Polymers	Blane
Canadian General Tower	Resilon
EVC	Corvic (homopolymers and copolymers)
	Welvic
EVC	Vipla
Hoechst/Hoechst Celanesee	Hostalit
Huls	Vestolit
Hydro Polymers	Hyvin
Laporte Ind.	Benvic IR
Laporte Ind.	LVM
Neste	Neste PVC
Neste Oy	Pekevic
Polimeros De Mexico	Irtavil
Schulman	Polyvin
Shell	Rovin
Solvay	Benvic IR
Vista	Vista Chemicals
Wacker Chemie	Vinnol

6 Material Properties
The discovery that plasticized PVC could be converted to an elastomeric-type material, by the addition of low molecular

weight plasticizers, was made many years ago (by Dr Waldo Semon in the 1920s) when attempts were made to dissolve the polymer. Polymeric plasticizers were used approximately 10 years later and the use of elastomers, as property modifiers, was introduced shortly afterwards. Then, the preferred elastomers were copolymers of acrylonitrile and 1,3 butadiene: these materials are known as nitrile elastomers (NBR) and such materials are still used with loadings of up to approximately 33phr. (Could, of course, use mixtures of high molecular weight elastomers and of low molecular weight plasticizers).

As there are many different plasticizers, which may be used in different amounts and/or in combination, then a wide range of plasticized compounds is possible. Such compounds will differ not only in flexibility and softness, but also will differ in other respects, for example, tensile strength, resilience and ease of flow. These latter properties can also be affected by the combination of different plasticizers and molecular weight resins.

Plasticizers are added to PVC compounds in order to confer flexibility and softness. The plasticizers most commonly employed are high boiling point esters of C_{8-10} alcohols: such as phthalates, phosphates and sebacates. Examples of common plasticizers are dioctyl phthalate (DOP, which is also known as 2-ethylhexyl phthalate), di-iso-octyl phthalates (DIOP) and, dialphanyl phthalate (DAP).

The term 'plasticized' (also spelt 'plasticised') simply means that there is plasticizer (plasticiser) present in the injection molding material but, it does not mean that there are no other additives mixed in with the polyvinyl chloride resin (PVC). Additives are essential in PVC technology as without them the plastic is useless as it cannot be successfully processed. Some additives enhance the properties of the base PVC and permit its use in a wider range of applications.

Examples of additives include:-

(i) heat stabilizers - often based on lead compounds, calcium/zinc mixtures or, on tin complexes. Epoxidized soya bean oil, at low concentrations, is used as a component of some stabilizer systems. It is important to ensure that the compound contains sufficient stabilizer so that reprocessing is practical at the desired level of regrind use;

(ii) lubricants - for example, both internal and external lubricants are necessary. Calcium stearate is an example of an internal lubricant whereas synthetic waxes and fatty acid esters are used as external lubricants. Lubricant packages are formulated specifically to suit a particular application/machine combination. The level of use is often kept low (below 1phr or, one part per hundred of resin) in order to reduce plate out problems and a loss of impact strength;

(iii) processing aids - which may be based on acrylic polymers; and,

(iv) fillers - to extend the scope of PVC, and/or to make it cheaper, fillers (such as china clay and whiting) are often used.

The ingredients of the PVC mix must be blended together and then, at some stage, the blend must be melt compounded. Control over the compounding stage is important as, for example, material flow and end-use properties, are effected by the level of mixing introduced during compounding.

The PVC material may be bought in as compound (which means that it has been melt compounded) or, it may be blended, in for example, a high speed mixer and then fed to the injection molding machine or, it may be blended, compounded and molded in-house. For this reason the material may be known by the suppliers name of the parent plastic or, it may be known by the name of a compound. In general, the injection molding of PVC compounds gives the best properties but the molding of dry blends can offer cost advantages, however, the throughput of dry blend must be sufficient to justify the installation costs. Most plasticized PVC compounds are based on PVC resins having a K value of between 65 to 70 with a porosity factor that is sufficiently high enough to rapidly absorb the plasticiser during mixing to produce a dry free flowing blend. Use of such compounds enables PPVC to be processed similar to other thermoplastics materials. To realize the potential of this material it is necessary to put in just sufficient work and heat so that the blend is fully gelled, or fused, but not so much that it is degraded. Such decomposition can be caused by over-heating or by excessive shear. Once decomposition starts it can spread very rapidly as, for example, one of the products of decomposition (hydrochloric acid or, HCL) catalyses further degradation. Being an acid it also readily attacks metals such as steel and causes pitting and corrosion; it also promotes rust as it strips protective layers away from metals such as mild steel. HCL fumes can also be an irritant to people and contact is best avoided.

Despite these problems PVC is widely used as it is relatively cheap and it is incredibly versatile; however, its inherent flame retardancy is often reduced by plasticizer addition. The chemical resistance is good and when correctly formulated, transparent injection moldings are readily obtained; this means using appropriate stabilizers, for example tin stabilizers and calcium/zinc complexes.

7 Ease of Flow

The ease of flow, of a PVC resin or polymer, is dependent on the molecular weight of the resin which is characterized for PVC by the 'K' value; the bigger the number the higher is the molecular weight and the more difficult the flow. For a particular K value the ease of flow may be dramatically altered by the use of additives, for example, by the use of lubricants or plasticiser. As PPVC contains plasticiser, which eases flow, high K value resins can be used for injection molding, for example, those with a DIN K value of between, approximately, 65 and 70.

MOLDFLOW DATA FOR HULS VESTOLIT

Grade.	Viscosity (Nsm^{-2}) at $1,000s^{-1}$ at the following temperatures			
	140°C (284°F)	160°C (320°F)	180°C (356°F)	200°C (392°F)
General Purpose grade (SP50)	300	145	68	32

Material Grade	Viscosity (Nsm^{-2}) at 180°C (428°F)			
	$100s^{-1}$	$1,000s^{-1}$	$10,000s^{-1}$	$100,000s^{-1}$
Vestolit (SP50)	270	68	18	4

Please note that the viscosity at $100,000s^{-1}$ shows a dramatic change in the flowability. However the mold filling speeds used for PPVC would seldom incur shear rates of $100,000s^{-1}$ due to degradation occurring.

8 Shrinkage

This is of the order of 0.01 to 0.02 in/in or mm/mm ie 1.0 to 2.0%. Shrinkage variation is due primarily to differences in the hardness value of the compound. Shrinkage values are usually lower across the flow direction: however, the actual value obtained is very dependent upon the molding conditions used.

9 Resistant to the Following

Water and salt solutions; dilute acids and alkalis have little effect at room temperature but, at elevated temperatures, some hydrolysis and extraction of the plasticizer may occur. Concentrated acids and alkalis hydrolyse plasticizers slowly when cold but, more rapidly when heated. Most organic liquids will extract plasticizers and cause compound hardening however, some organic liquids also cause softening. For example, aromatic and chlorinated hydrocarbons, aliphatic and aromatic nitro compounds, ketones, aliphatic and cyclic ethers. The effect of stabilizers, lubricants and pigments on the chemical resistance of PVC compounds is less marked than the effect of plasticizers and copolymers; the type and amount of which, will affect the chemical resistance.

To get improved chemical resistance replace a monomeric plasticizer, such as dioctyl phthalate (DOP), with a polymeric type, for example, polypropylene adipate (PPA) or polypropylene sebacate (PPS). PPA is useful for oil and fuel contact and PPS resist soaps and detergents.

PPVC is less permeable than PE; the permeability increases with plasticizer content.

10 Not Resistant to

Acids, alkalis, oxidizing agents. More readily attacked by solvents than UPVC. Attacked by solvents such as tetrahydrofuran, cyclohexanone, methyl ethyl ketone (MEK), dimethylformanide. Plasticized PVC is also readily swollen aromatic and chlorinated hydrocarbons, nitrocarbons, acetic anhydride, aniline and acetone.

11 Material Detection or Identification

Density depends on the formulation: in general however, the density of most formulation will be greater than $1 gcm^{-3}$ and so the material (solid, non-filled material) will sink in water. As the density of most plasticizers is less than that of PVC (which is $1.4 gcm^{-3}$) then, moldings based on only PVC and plasticizer, will float in saturated, magnesium chloride. If high levels of filler are present then, the density will be increased. (The presence of inorganic fillers can be shown by burning off the organic portion of the compound: this is mainly the PVC and the plasticizer). Stabilizers, particularly the metal compounds, are also partially responsible for an increase in the density.

The natural color of the material based on PVC and plasticizer, is transparent but such compounds are unusual. Most other additives will worsen transparency.

When this material is heated in a flame then, its behavior depends tremendously on the compound composition, for example, the amount and type of plasticizer. Most plasticizers burn easily and so their presence worsens flame resistance and allows the compound to burn more easily. However, some are naturally flame resistant (the phosphates, for example, trixylyl phosphate or TXP) and compounds based on such materials will not burn easily. Because of their high price, phosphates are only used where flame resistance is essential as they are toxic and, give high cold flex temperatures. In general therefore, plasticized PVC compounds burn when heated in a flame to give a yellow, sooty flame: a black residue is formed and often the flame has a green tip.

When heated in the absence of a flame then, low molecular weight plasticizers can be distilled off and, analyzed if required. A better way of doing this is to solvent extract a known quantity of the injection molding material with a solvent, remove the solvent from the extract and then analyze the liquid plasticizer. A useful test, for plasticizer characterisation, is boiling point.

PPVC compounds can be cut very easily with a knife and can be scratched easily with a knife or fingernail

This material will dissolve in solvents such as tetrahydrofuran, cyclohexanone, methyl ethyl ketone (MEK), dimethylformanide. Plasticized PVC is also readily swollen aromatic and chlorinated hydrocarbons, nitrocarbons, acetic anhydride, aniline and acetone.

This material will not dissolve in solvents such as aliphatic hydrocarbons but such solvents can leach out plasticizers. This material will swell in solvents such as aromatic and chlorinated hydrocarbons, nitrocarbons, acetic anhydride, aniline and acetone.

12 Coloring

As the natural color of the material can be clear a wide color range is possible, both transparent and opaque. It is sold in both compounded colors and as natural material for coloring on the injection molding machine by techniques such as dry coloring, masterbatching and liquid coloring. With PPVC, only pigments should be used because of color bleeding, or leaching problems, with dyestuffs. Certain types of pigments can catalyze (progress the rate of) decomposition and/or cause color changes to occur to the original hue; metal powders or oxides fall into this catagory.

It is much more difficult to color match opaque PPVC compounds than clear (crystal) ones, due to color variation in the base materials. Depending upon the nature of the PPVC compound (i.e. granular or powder form) a binder may be required to adhere the colorant to the polymer during the mixing stage. A typical binder would be DOP (i.e. dioctyl phthalate) which is added to the mix at a concentration level of 0.1%. To achieve the desired properties, many PVC compounds may include a plasticizer system other than DOP, but the inclusion of such a small amount of DOP will not significantly alter the end properties of the molded component. Typical dry colorant loadings of 0.05 to 2.5% are used: the actual percentage depends upon whether a transparent or opaque color is required.

PVC based masterbatches are used to achieve good color compatability, but their usage is relatively limited. Universal masterbatches are used but due to their partial compatability with the molding compound, not all of them are suitable so the manufacturer's advice should be sought before use. Masterbatch additions, for both PVC-based and universal types, are between 1 to 5% thus giving typical pigment loadings of 0.5 to 2% in the final product (depending upon the denseness of color required).

Liquid colors can also be used, but their addition must be controlled so as not to create changes in the properties of the molding compound, as a result of the liquid carrier.

13 Materials and Components

Will absorb less than 0.02% water in 24 hours at room temperature. This means that drying is not normally necessary. If it is, dry in a hot air oven at 65°C (149°F) for 2 hours or, in a desiccant dryer for 1 to 1.5 hours at 70°C (158°F). When drying PPVC compounds, strict temperature and time control must be observed so as not to cause too great a loss of plasticizer (especially when heating highly plasticized compounds or those containing a highly volatile type of plasticizer). Such losses effect the processing characteristics and the properties of the PVC compound. Prolonged heating can also effect the thermal stability during processing due to some loss of stabilizer. Do not pre-dry PPVC with other materials as they may become contaminated by the plasticizer evaporating from the PPVC compound.

When a desiccant dryer is used for PPVC compounds, it is essential that a plasticizer trap is fitted within the dryer so as to collect the vaporized plasticizer and prevent it from contaminating the desiccant bed. Due to the problems relating to liquid plasticizers their replacement with various rubbers i.e. NBR, is being actively pursued.

To produce a PPVC dry blend, the PVC polymer and other ingredients (with the exception of the plasticizer and lubricant) are fed into the chamber of a high speed mixer and blended at high speed until a temperature of between 90 to 95°C (194 to 203°F) is reached. The plasticizer is then added to the mix and when 110 to 125°C (230 to 257°F) is reached, lubricant may be added; the resultant mix is dumped into a cooler/blender. A large cooler/blender, for example, a ribbon blender, may be used. This cools the mix, improve the batch-to-batch consistency and also reduce the electrostatic charges on the dry blend. The dry blend may then be used as it is or, pelletized using an extruder.

Pelleting improves dispersion but adds to the cost and

reduces the thermal stability of the material. The addition of rubber instead of a plasticizer requires minor modifications to the above process, i.e. the rubber is added with all the other ingredients. Use of PVC dry blends is increasing within the injection molding industry as different types of PVC compounds can be readily developed to suit the end application of the component and the ease with which the material can be molded.

To improve the batch-to-batch consistency of a dry blend, the additives used in the dry blend (e.g. stabilizers, lubricants) can now be incorporated in pre-weighed sachets. These are known as 'one shot' or 'one pack' sachets; depending upon the number of ingredients used, the level of addition can be as high as 5 phr. This technique enables the molder to incorporate the correct percentage of additive easily and eliminates the hazards associated with the handling and weighing out of the individual ingredients.

To minimize the presence of dust, which is often associated with PVC dry blends, additives are now being supplied in a granular, flake or spaghetti form. Although the dispersion characteristics of the latter forms are not as good as when the additive is a powder, they are preferred by the processors due to the safety aspect.

14 Mold and Gate Considerations

A suggested mold temperature is 40°C. If a melt temperature of 200°C is used then the maximum amount of heat that needs to be removed for each gram of shot is 250 Joules(J). Over the temperature range quoted, the specific heat is approximately 1564 J/kgK.

Typical runner sizes are 5 to 9mm (0.197 to 0.354in). Runners, like sprues, are usually short in length and generous in diameter as this reduces pressure loss and permits the application of adequate follow up pressure. However, if they are made too large then excessively long cycles, and large material losses, result. If they are made too small then the large amount of pressure which is lost is transferred into heat (pressure losses of up to 1000psi (6.9MNm^{-2}) for every inch of runner length can be encountered on runner sections of less than 5mm (0.197in)). This heat will show up in the regions where the material is being sheared the most, i.e. in the gate regions. Such local temperature rises can be very high and can lead to severe material degradation. The feed system, e.g. the gate, should be sized so that the maximum shear rate is less than 30,000sec^{-1}.

It is suggested that the feed system be laid out as follows. The sprues should be as short as possible, of full round cross-section with a minimum diameter of 6mm (0.236in), at the nozzle entry, and of conical shape with an included angle of 5 degrees. Runners should also be of full round cross-section, they should incorporate cold slug wells and be connected to the sprue and gates via a rounded radius; a diameter of 7mm (0.276in) is suggested. Gates should also be connected to the molding by a rounded radius, short lands (0.7 to 1.0mm; 0.028 to 0.039in) should be used and smooth cross-sectional variations are essential.

The importance of a cold slug well is often overlooked; its inclusion prevents semi-solid material from entering the mold cavity and affecting the surface finish and strength of the product. Semi-solid material can also cause gate blockages, on a multi-impression mold, thus causing the other cavities to become overpacked.

For multi-impression molds, a balanced runner design is preferred; generous radii should be incorporated at each runner junction so as to prevent excessive shear heating to the material during mold filling and/or, causing unnecessary pressure drops to occur.

When selecting the gating position, ensure that there is a continuous flow of material at all points of the flow path with no stagnation at sharp corners; avoid regions of compression and decompression and keep the flow path very streamlined.

The following types of gate are often used, pin-point, fan, tab, edge, submarine, sprue and winkle. Pin-point and submarine types are used for various sized moldings where the smallest recommended diameter is 1mm (0.039in). If a submarine gate is to be used, then keep the entrance angle to the gate as wide as possible (i.e. up to 45°). Fan gating is used for thick sections as their use prevents weld lines and allows moderately fast mold filling speeds to be used; the thickness should not be below 1mm (0.039in). In general, gate depths of 0.7t to 1.0t, where t is the component wall thickness, are recommended. For moldings above 56g/2oz in weight it is not uncommon to use gate widths of 4 to 7mm (0.160 to 0.276in).

In general, a restricted or an impinged gate creates more turbulent melt flow thus impairing the surface finish of the molding. To improve the surface gloss, it is preferable to use a restriction in the runner side of the gate, rather than in the gate itself, so as to induce more frictional heat to the material.

Draft angles must be between 0.5 to 1 degree; ensure that there is adequate venting in the cavity. Typical vent sizes are 0.03 to 0.05mm (0.001 to 0.002in) deep by 8mm (0.315in) wide for surface vents or, use 0.03 to 0.05mm (0.001 to 0.002in) clearances around each ejector pin.

Molds which suffer from an excess of plasticizer condensate, or from regular blocking of the vents, are indicating that the mold filling speed is too fast.

Construct the mold from stainless steel with a minimum chromium content of 13% (preferably 16%), the Rockwell C hardness must be at least 55. A full range of surface finishes is possible from high gloss to textured on PPVC moldings. Hardened steel molds may be protected by electroless nickel plating followed by chrome plating. To give additional protection, surface treatments such as titanium nitriding or ion implantation are also applied.

Certain types of hot sprue bushing can be used for PPVC. However careful consideration needs to be given to the relative thermal stability of the molding compound (i.e. the type and amount of stabilizer system used) and the anticipated cycle time, before deciding which system to use. In general, hot runner molds are not recommended even though they have been successfully run. If attempting to use a runnerless design, then the following criteria must be met: the runner channels must be designed to allow the material an unrestricted flow path with no material hold-up areas, there must be uniform heat control throughout the runnerless system and the split type manifold should not be used if it can be avoided.

After completing a production run the mold surfaces should be washed scrupulously with a mild alkaline solution such as sodium nitrate. Alternatively dissolve 1 kilo (2.2lbs) of soda crystals in 1 liter (2.11 U.S. pints) of water. Use the prepared solution to wash the mold surfaces. After drying, the mold surfaces should be sprayed with an oil or silicone base spray.

15 Flow Path: Wall Thickness Ratio

For general purpose grades this is approximately 200 to 250:1 for a 1mm (0.039in) wall section; can reach 300:1 for easier flowing (highly plasticized) grades.

16 Projected Area

Usually 1.5 to 3tsi. (23 to 47MNm^{-2}) is sufficient. The actual value is dependent upon the type and amount (phr) of plasticizer used within the PVC compound.

17 Cylinder Equipment

It is recommended that the screw design be as follows. The screw is not usually fitted with a standard back-flow valve (of the sliding ring type) due to the risk of stagnation and subsequent degradation. However, due to the advantages found when a back flow valve is used (i.e. an improvement of cycle reproducibility and a reduction in the back flow of the melt) modified designs of the sliding ring type have been developed to suit PPVC.

A typical screw design has an L:D ratio of 18 to 24:1; shorter L:D ratios of 16:1 are successfully used but, for dry blends, the higher ratio is preferred. Barrier design screws are also used when processing PVC dry blends to provide better homogeneity of melt. When using these types of screws, the amount of heat developed can often be excessive and therefore it is usual to provide screw cooling – particularly in the tip region. The longest section of the screw is the compression zone (usually 0.4L for a typical screw); the feed and metering zones are 0.3L and 0.3L respectively. The compression ratio ranges from 1.2 to 2.1:1 depending upon the type of compound and size of molding machine used. The depth of flight in the metering section is important so as to prevent excessive shear being imparted to the melt; typical values are given below:

Flight depth		Diameter of screw	
mm	(in)	mm	(in)
3	(0.118)	35	(1.38)
4	(0.156)	50	(1.97)
5	(0.196)	60	(2.36)
5	(0.196)	75	(2.95)
7	(0.276)	100	(3.94)

The design of the screw tip is such that the inclusive angle is usually between 25 to 30 degrees; the surface of the tip should be highly polished or of the spiral (self wiping) design. The optimum clearance between the screw tip and the nozzle, when the screw is in it's full forward position, should be 0.7 to 1.8mm (0.028 to 0.072in). This small gap ensures that a only a very small amount of PVC remains in the nozzle after each injection cycle. The clearance between the screw flight and barrel wall should be in the region of 0.012 to 0.2mm (0.005 to 0.008in).

The components of the barrel assembly which are in direct contact with hot PVC, must be able to resist the corrosive hydrochloric acid (HCL). To protect against such corrosion the screw, barrel and back flow valve are often produced from stainless steels containing a high chromium content. Chrome plating, in conjunction with nitriding, is often used for bi-metallic barrels; fully hardened screws are produced from other types of steels. Screws possessing a deep nitrided surface offer good resistance to HCL attack. The cylinder assembly should be designed so that all sharp corners, abrupt changes in cross-section and direction are avoided, so as to prevent the material from becoming trapped in a nil flow area thus causing it to degrade. It is usual to keep a separate screw and barrel assembly for PVC compounds.

An open nozzle having zero, or a very gradual, taper with a perfectly smooth bore should be used. A reverse taper geometry with a diameter of at least 5mm (0.196in) is recommended: keep the length of the nozzle as short as possible i.e. 50 to 100mm (1.97 to 3.94in). The exit diameter of the nozzle should be smaller than the entrance diameter of the sprue bush by 1mm (0.039in). Keep the orifice land at the front of the nozzle to a minimum i.e 4 to 6mm (0.156 to 0.236in). Accurate temperature control of the nozzle is vital to avoid thermal degradation. 'Shut off' nozzles are not recommended for use with this material due to the problem of degradation. Decompression or suck back may be used; however, it is not recommended so as to prevent oxidation occurring to the material.

A correctly designed hopper is essential when processing dry blends: the angle of the hopper walls should be between 45 to 60 degrees. Angles of less than 45 degrees can cause bridging problems. If bridging does occur the use of an agitator can overcome such a problem.

18 Screw Cushion

On smaller machines this is often approximately 3mm (0.118in); on larger machines 9mm (0.354in) is suggested. However, because of the risks of decomposition, or degradation, when processing PVC, it is suggested that the screw cushion be kept as small as is practicable. The optimum clearance between the screw tip and the nozzle, when the screw is in it's full forward position, should be 0.7 to 1.8mm (0.028 to 0.072in). This small gap ensures that a only a very small amount of PVC remains in the nozzle after each injection cycle.

19 Shot Capacity

Shot capacities of 75% and above are recommended; if the screw has been designed for PPVC then up to 95% capacity can be used. The smaller the shot size, compared to the total capacity of the barrel, the longer is the material residence time and therefore the greater the risk of thermally degrading the material.

20 Melt Temperature

This is as measured in the nozzle or by an air shot technique; 175°C to 200°C (329°F to 392°F) is suggested. At the correct melt temperature, the PPVC purgings are smooth and glossy. If the purgings have a bumpy or lumpy appearance, then the material is non-homogeneous (i.e. not sufficiently thermally plasticized) – indicating that the set temperatures are too low. If the purgings have a bubbly/frothy appearance, and give off an excessive amount of smoke, then the material is too hot.

21 Barrel Residence Time

At a temperature of 200°C (392°F) the maximum residence time in the barrel should not exceed 5 minutes. When the temperature is 205°C (401°F) then the maximum residence time in the barrel should not exceed 3 minutes.

22 Temperature Settings

Please note that it is the melt temperature which is important; those temperatures given in the table are only suggested, initial settings. The temperature of the hydraulic oil, the hopper block and the material temperature should also not vary excessively. Mold temperature control is extremely important and therefore these temperatures must be checked regularly.

The temperature controllers should be initially set towards the lower end of the temperature settings range and then gradually increased to achieve the desired melt temperature. When using dry blends, it is important to use a temperature profile which increases along the barrel. This ensures that entrapped air is forced backwards, along the screw and out of the hopper. The effect of entrapped air can often mislead the processor into thinking that the melt temperature being used is too high – due to the bubbly/frothy appearance of the purgings.

Zone No.	Location	Temperature in °C FROM	TO	Temperature in °F FROM	TO
0	Hopper throat	30	50	86	122
1	Barrel rear	140	160	284	320
2	Barrel middle	150	170	302	338
3	Barrel middle	160	180	320	356
4	Barrel front	170	190	338	374
5	Nozzle	180	205	356	401
6	Mold	30	50	86	122

23 Injection Speed

The mold filling speeds used with PPVC need to be relatively slow as excessive shear can cause the material to degrade or its use will effect the surface finish of the molding (i.e. splash/mica marks or surface delamination). Profiled mold filling speeds are often used to produce components which require a good surface gloss.

24 Injection Pressure

The machine should be capable of giving up to:
> first stage, up to 1500bar; 150MNm^{-2}; 21,750psi.
> second stage, up to 750bar; 75MNm^{-2}; 10,875psi.

25 Screw Rotational Speed

Should be adjusted to suit the molding cycle but do not use a screw surface speed of greater than 0.15 to 0.2ms^{-1} (0.49 to 0.66ft/sec). If a screw of 40mm (1.57in) diameter is being used then the rpm should not exceed 95. Similarly, the maximum diameter of a screw when rotated at 200 rpm in order to obtain a screw surface speed of 0.15 to 0.2ms-1 (0.49 to 0.66ft/sec), is 19mm (0.75in).

26 Back Pressure

Up to 80bar; 8MNm^{-2}; 1160psi should be available. Back pressure should be kept as low as possible: usually a nominal value of 5 bar (0.5MNm^{-2}, 72.5psi) is used so as to achieve consistent shot weights. May need to be increased, to assist dispersion, when using dry blends or colorants but keep as low as possible.

27 Shutting Down

As the thermal stability of PPVC is poor then the shutting down procedure is very important. Ensure that all traces of the PPVC have been removed from the barrel by purging thoroughly with another thermoplastic material which is more thermally stable and does not react with PPVC (for example PMMA, PP, LDPE, HDPE or GPPS). Usual procedure for shutting down, for short periods, is to switch off the barrel heaters and purge the barrel clean.

NEVER MIX POLYACETAL (POM) and PPVC TOGETHER IN THE SAME BARREL as the resultant vigorous chemical reaction can either cause considerable damage to the molding machine or, serious injury to molding shop personnel. If PPVC is to follow POM (or visa versa) then thoroughly purge the barrel clean with one of the materials stated above first. In some cases it is preferable to strip down and clean the screw and barrel assembly. It is common practice to keep a separate screw and barrel assembly solely for PPVC.

28 Reprocessing

When this material is reclaimed it is suggested that up to 20% be blended with the virgin material. The end usage of the molded component dictates the level of regrind that can be used. What must be borne in mind, is that each time PPVC is reprocessed it's resistance to thermal degradation and flowability is considerably reduced. Instances have occurred where moldings fail while in service due to the abuse of regrind usage. Problems such as these occur not only because of the percentage of regrind being incorporated, but also as a result of using regrind which has been degraded as a consequence of being reprocessed too many times.

29 Finishing

PPVC can be welded by various techniques such as high, or radio frequency, heated tool and solvent welding. The effectiveness of the joining method is dependent upon, for example, the amount of plasticizer and the formulation of the compound. Depending upon the hardness, PPVC can be machined using conventional machining techniques; keep the surfaces being cut well cooled. Hand degating (using scalpels, side cutters, etc.) is common; automatic degating is used, e.g. with compounds containing lower concentrations of plasticizer.

30 Other Comments

Vinyl chloride (VC) polymers may be made by emulsion (giving PVC-E), suspension (PVC-S) and bulk, also known as mass, polymerization (PVC-M). As VC is toxic the level of monomer left in the plastic must be kept very low ,for example, less than 1 part per million (<1ppm). PPVC is usually based on PVC-S or PVC-M.

31 Typical Components

Typical components include washers, grommets, electrical cable ends, footwear and heel tags, watch straps, electrical shields, plugs, automobile arm rests, knobs, metal reinforced steering wheels and components associated with the medical industry.

PVC is widely used in packaging: such packaging is often disposable. A common way of disposing of waste/rubbish is to burn it in incinerators. PVC has for many years been under suspicion as being potentially harmful to health, and to the environment, as being a precursor in the formation of dioxins and furans. Incineration studies do not support this view and some people therefore believe that incineration is a safe and viable disposal option. Operating conditions during burning, rather than the material being burnt, appear to have a more significant effect on toxic gas emissions. Do not operate incinerators at temperatures below 750°C (1400°F) as otherwise high dioxin levels result. To burn any plastic does, however, seem silly and particularly so for PVC as, HCL is formed. Waste PVC can be recycled and the trend towards recycling such materials is growing.

32 Unit Price

If the price of a natural, unmodified grade of GPPS is taken as one then the relative price of this molding material is approximately:

General Purpose grade of PPVC   0.8 to 0.9
Medical grade of PPVC   1.2 to 1.5

The relative price compared to GPPS can also be found in Table 7.

UNPLASTICISED POLYVINYL CHLORIDE

1 Common Name

Unplasticised polyvinyl chloride.

2 Abbreviation(s)

UPVC ; RPVC ; PVC-U.

3 Alternative Names

Unplasticised PVC; unplasticized polyvinylchloride or unplasticized PVC ; rigid PVC ; hard PVC. PVC may also be referred to as poly(vinyl chloride) or as poly(monochloroethylene).

4 Some Suppliers / 5 Trade Names or Trade Marks

Some Suppliers	Trade Names or Trade Marks
Atochem	Lacqvyl
BASF	Vinoflex/Vinidur
EVC	Corvic (polymers and copolymers)
EVC	Prima (PVC alloys)
Goodrich BF Chemical	Geon
Hoechst	Hostalit
Hoechst	Hostalit
Huls	Vestolit
Laporte Ind.	Benvic IR
Neste	Neste PVC
Norsk Hydro Polymers	Hyvin
Schulman	Polyvin
Shell	Rovin
Solvay	Benvic IR
Wacker Chemie	Vinnol

6 Material Properties

The term 'unplasticized' (also spelt 'unplasticised') means that there is no plasticizer present in the injection molding material but, it does not mean that there are no additives mixed in with the polyvinyl chloride plastic or, resin (PVC). Additives are essential in PVC technology as without them the plastic is useless as it cannot be processed. Some additives enhance the properties of the base PVC and permit its use in a wider range of applications. Examples of such additives include heat stabilizers (often based on lead or tin), lubricants (for example, butyl stearate), processing aids (which may be based on acrylic polymers), impact modifiers (based on plastics such as ABS or MBS or, on rubbers, or elastomers, such as NBR) and fillers (such as china clay). The PVC material may be bought in as compound (which means that it has been melt compounded) or, it may be blended, in for example, a high speed mixer and then fed to the injection molding machine or, it may be blended, compounded and molded in-house. For this reason the material may be known by the suppliers name of the parent plastic or, it may be known by the name of a compound. In general, the injection molding of PVC compounds gives the best properties but the molding of dry blends can offer cost advantages, however, the throughput of dry blend must be sufficient to justify the installation and costs.

To realise the potential of this material it is necessary to put in just sufficient work and heat so that the blend is fully gelled, or fused, but not so much that it is degraded. However, if sufficiently low 'K' value resins are used (for example 50 to 55) then in many respects the PVC compounds, based on these low 'K' value resins, may be treated similar to other thermoplastics materials. The important factor is to process the material at the correct melt temperature.

Such decomposition can be caused by over-heating or by excessive shear. Once decomposition starts then it can spread very rapidly because, for example, one of the products of decomposition (hydrochloric acid or, HCL) catalyses further degradation. Being an acid it also readily attacks metals such as steel and causes pitting and corrosion; it also promotes rust as it strips protective layers away from metals such as mild steel. The effects on human beings are also not beneficial; in fact, they are very harmful.

Despite these problems the material is widely used as it is relatively cheap, inherently flame retardant, and is strong and stiff. The chemical resistance is good, for example it has good oil resistance, and yet it is possible to join it by solvent welding. When correctly formulated, transparent injection moldings are possible; this means using appropriate stabilizers, for example tin stabilizers, and matching the refractive index of, for example, the impact modifier and the PVC.

As with many other plastics materials PVC is now being blended, or alloyed, with other plastics or rubbers. This permits the development of new materials with different properties and low cost. One such plastic blend is that between PVC and ABS which gives a flame retardant ABS type material. When blended with appropriate grades of the rubber known as NBR then it is possible to produce a thermoplastic elastomer or TPE. The properties of some of these TPE materials can be further enhanced by ionic crosslinking.

7 Ease of Flow

A relatively stiff flow material with a limited temperature processing range. The ease of flow is dependent on the molecular weight which is characterized for PVC by the 'K' value; the bigger the number the higher is the molecular weight and the more difficult the flow. For a particular 'K' value the ease of flow may be dramatically altered by the use of additives, for example, by the use of lubricants. In general however, relatively low 'K' value resins are used for injection molding. Up to fairly recently the typical DIN 'K' value used was between, approximately, 55 and 62 (these figures correspond to ISO viscosity numbers of, approximately 74 and 95. The inherent viscosity (ASTM) is approximately 1/100 of the viscosity number); now that the 'K' value range has been reduced to 50 to 60 these lower molecular weight materials can be processed much more easily.

MOLDFLOW DATA OF EVC PRIMA

Material	Viscosity (Nsm^{-2}) at 1,000s^{-1} at the following temperatures			
	160°C (320°F)	180°C (356°F)	200°C (392°F)	220°C (428°F)
Easy flow grade (RI5022)	593	302	154	79
Medium flow grade (RI5039)	753	431	246	141
Stiff flow grade (RI5070)	1071	645	389	235
High Heat Resistant grade (KR2556)	423	253	151	90

Material Grade	Viscosity (Nsm^{-2}) at 200°C (464°F)			
	100s^{-1}	1,000s^{-1}	10,000s^{-1}	100,000s^{-1}
Prima (RI5022)	493	154	48	15
Prima (RI5039)	1032	246	59	14
Prima (RI5070)	2174	389	70	13

Please note that the viscosity values given at 100,000s^{-1} shows a change in the flowability of each grade i.e. the stiff flow grade becomes slightly easier flowing than the easy flow grade. However the mold filling speeds used for UPVC would seldom incur shear rates of 100,000s^{-1} due to degradation occurring.

8 Shrinkage

This is of the order of 0.002 to 0.006in/in or mm/mm i.e. 0.2 to 0.6%. The low shrinkage value enables close toleranced moldings to be readily produced.

9 Resistant to the Following

Water, salt solutions, oxidising agents (for example, hydrogen peroxide), reducing agents, hypochlorite solutions, aliphatic hydrocarbons, detergents, non-oxidising acids and concentrated alkalis. UPVC is resistant to most oils, fats, alcohols and petrols. Highly resistant to strong acids, for example, any concentration of hydrochloric acid at temperatures up to 60°C (140°F); any concentration, up to 90%, of sulfuric acid at temperatures up to 60°C (140°F); cold, 50% nitric acid does not attack UPVC. Chemical and weathering resistance is good but is generally made worse by the addition of impact modifiers. Homopolymers of PVC are more resistant to chemical attack than copolymers. UPVC compounds are not recommended for environmental and continuous use above 60°C (140°F).

10 Not Resistant to

Concentrated, oxidising acids such as sulfuric, nitric and chromic acids which cause decomposition; the rate of decomposition may be accelerated in the presence of metals, for example, zinc and iron. Attacked by bromine and fluorine even at room temperature. Unsuitable for use in contact with aromatic and chlorinated hydrocarbons, ketones, nitro-compounds, esters and cyclic ethers; these penetrate the PVC and cause marked swelling. Can have a high impact strength but can be very notch sensitive. Methylene chloride can be used to detect an under-gelled compound.

11 Material Detection or Identification

PVC is a substantially amorphous, thermoplastic material which does not exhibit a sharp melting point. With a density of approximately 1.4gcm^{-3} (0.81 oz/cu.in.) the material (solid, non-filled material) will sink in both water and in saturated, magnesium chloride. The density of commercial compounds may range from 1.34 to 1.45gcm^{-3} (0.78 to 0.84 oz/cu.in.). The natural color of the material is clear when it is unfilled and stabilized with appropriate stabilizers, for example, tin stabilizers. If the material is stabilized with lead stabilizers, and /or filled, then the natural color is usually an off-white. In either case this means that a wide color range is possible. Many compounds have a hardness, on the Shore D scale, of about 80 and a Vicat softening point (50N) from 65 to 90°C (149 to 194°F).

When this material is heated in a flame then it softens slightly and chars; self extinguishing but if heated strongly then a dirty yellow flame is seen together with a lot of smoke and an acrid smell; edges of the flame may be tinged green. When heated in the absence of a flame then this material will give off first HCL and then various hydrocarbons. If a piece of this material is heated on a copper wire then the flame is colored green.

Can be cut easily with a knife and can also be scratched relatively easily with a knife or fingernail. Has a low softening point – well below that of boiling water.

This material will dissolve in solvents such as tetrahydrofuran (THF), cyclohexanone, methyl ethyl ketone (MEK) and dimethylformanide. Also by solvent mixtures based on, for example, acetone or benzene. (Plasticized PVC is also readily swollen by aromatic and chlorinated hydrocarbons, nitrocarbons, acetic anhydride, aniline and acetone). This material will not dissolve in solvents such as aliphatic hydrocarbons (such solvents can leach out plasticizers in plasticized PVC). UPVC will swell in solvents such as aromatic and chlorinated hydrocarbons, nitrocarbons, acetic anhydride, aniline and acetone.

12 Coloring

As the natural color of the material can be clear then a wide color range is possible; this includes both transparent and opaque colors. Sold in both compounded colors and as natural material for coloring on the injection molding machine by techniques such as dry coloring, masterbatching and liquid coloring. When coloring UPVC both dyestuffs and pigments can be used. However certain types of pigments can catalyse, or progress the rate of, decomposition; metal powders or oxides fall into this category. The color matching of opaque UPVC compounds can sometimes be difficult due to color variation in the base material, occurring as a result of the types or quantities of additives used. Depending upon the nature of the UPVC compound (i.e. granular or powder form) a binder may be required to allow the colorant to adhere to the polymer during the mixing stage. A typical binder would be DOP (i.e. dioctyl phthalate) which is added to the mix at a concentration level of 0.1%. Typical dry colorant loadings of 0.05 to 2.5% are used: the actual percentage is dependent upon whether a transparent or opaque color is required.

The most commonly used type of masterbatch is PVC based, however only some universal masterbatches are suitable and therefore advice should be sought from the manufacturers before use. Masterbatch additions of between 2 to 4% are used giving typical pigment loadings of 1 to 2% in the final product, depending upon the denseness of color required. Liquid colors can also be used, however the addition level must be controlled so as not to create changes in the properties of the molding compound, as a result of the liquid carrier.

13 Mold and Gate Considerations

A suggested mold temperature is 40°C (104°F). If a melt temperature of 200°C (392°F) is used then the maximum amount of heat that needs to be removed for each gram of shot is 291 Joules(J). Over the temperature range quoted, the specific heat is approximately 1820 J/kgK.

Typical runner sizes are 6 to 10mm (0.236 to 0.394in). Runners, like sprues, are usually short in length and generous in diameter as this reduces pressure loss and permits the application of adequate follow up pressure. However if they are made too large then excessively long cycles and large material losses result. If they are made too small then the large amount of pressure which is lost is transferred into heat. This heat will show up in the regions where the material is being sheared the most, that is, in the gate regions. Such local temperature rises can be very high and can lead to severe material degradation. The feed system, for example the gate, should be sized so that the maximum shear rate is less than 20,000sec^{-1}. It is suggested that the feed system be laid out as follows:-

the sprues should be as short as possible, of full round cross-section with a minimum diameter of 6mm (0.236in), preferably 8mm (0.315in), at the nozzle entry, and of conical shape with an included angle of 5 degrees;

runners should also be of full round cross-section which incorporate cold slug wells and connected to the sprue and gates via a radius, a diameter of 8mm (0.315in) is suggested;

gates should also be connected to the molding by a radius, short lands and smooth cross-sectional variations are essential.

The importance of a cold slug well is often overlooked as its inclusion prevents semi-solid material from entering into the mold cavity. The presence of such material considerably affects the surface finish and strength of the product. For multi-impression molds a balanced runner design is preferred and generous radii should be incorporated at each runner junction so as to prevent unnecessary pressure drops.

When selecting the gating position ensure that there is a continuous flow of material at all points of the flow path with no stagnation at sharp corners or, at chips and scratches in any metal surface; avoid regions of compression and decompression and keep the flow path very streamlined.

The following types of gate are often used, pin-point, fan, tab, edge and submarine. Pin-point and submarine types are used for small moldings and the smallest diameter is 1mm (0.039in). If a submarine gate is to be used then keep the entrance angle to the gate as wide as possible (i.e. up to 45°). Fan gating is used for thick sections as their use presents prevents weld lines and allows moderately fast mold filling speeds to be used; the thickness should not be below 1mm (0.039in). In general, gate depths of 0.7t to 1.0t, where t is the component wall thickness, are recommended.

In general a restricted gate creates more turbulent melt flow thus impairing the surface finish of the molding.

Draft angles must be between 0.8 to 2 degrees (per side) and ensure that there is adequate venting in the cavity. Typical vent sizes are 0.03 to 0.05mm (0.001 to 0.002in) deep by 6mm (0.236in) wide for surface vents or 0.03 to 0.05mm (0.001 to 0.002) clearances around each ejector pin.

Construct the mold from stainless steel with a minimum chromium content of 13%, preferably 16%; the Rockwell C hardness must be at least 55. A full range of surface finishes is possible, on UPVC moldings, from high gloss to textured. Hardened steel molds may be protected by chrome plating.

The insulated runner type of runnerless mold and certain types of hot sprue bushings can be used for UPVC, however careful consideration needs to be given to the relative thermal stability of the molding compound (i.e. the type of stabilizer system used) and the anticipated cycle time at which the mold is expected to run at before deciding which approach to opt for.

Where pipe fittings are being injection molded, ensure that the machine can generate sufficient opening force as the contracting material can bind the system together thus preventing the withdrawal of the side cores. Similarly components possessing undercuts, such as internal thread forms, may sometimes become jammed in the mold preventing the unscrewing mechanism to operate. To alleviate the problem the cores should be lubricated and the moldings removed whilst still in a hot condition.

After completing a production run the mold surfaces should be washed scrupulously with a mild alkaline solution such as sodium nitrate. After drying, the mold surfaces should be sprayed with an oil or silicone base spray. Mold surface temperatures and size of ejectors have a considerable influence on the cycle time with respect to deep draw components, such as computer housings etc..

Use the maximum area of ejection as possible and wherever possible, keep the mold surface temperature to below 40°C (104°F).

14 Flow Path: Wall Thickness Ratio

For general purpose grades this is approximately 80 to 100:1 for a 1mm (0.039in) wall section; can reach 200:1 for easier flowing grades.

15 Projected Area

Usually 2 to 3tsi. (31 to 47MNm^{-2}) is sufficient.

16 Materials and Components Handling

Will absorb less than 0.02% water in 24 hours at room temperature. This means that drying is not normally necessary. If it is, dry in a hot air oven for hours at 65°C (149°F) for 3 hours, in a dessicant dryer for 1 to 1^1/$_2$ hours at 80°C (176°F). UPVC is either used in the form of granules or as a dry blend. To produce a UPVC dry blend the PVC polymer is added together with other ingredients into the chamber of a high speed mixer. This mixture is blended at high speed until a temperature of between 110 to 125°C (230 to 257°F) is reached. Within this temperature range a lubricant and/or impact modifier may be added, then the resultant mix is dumped into a cooler/blender. A large ribbon blender is used to cool the mix, improve the batch consistency and also reduce the electrostatic charges on the dry blend. The dry blend may then be used directly or pelletized by using an extruder. Pelletizing improves dispersion but adds to the cost and reduces the thermal stability of the material. Use of PVC dry blends is increasing within the injection molding industry as different types of PVC compounds can be readily developed to suit the end application of the component and the ease at which the material can be molded.

To improve the batch to batch consistency of a dry blend, the additives used in the dry blend (e.g. stabilizers, lubricants, processing aids, impact modifiers and fillers) can now be incorporated in preweighed sachets. These are known as 'one shot' or 'one pack' sachets where, depending upon the number of ingredients used, the level of addition can be as high as 10 phr. This technique enables the molder to incorporate the correct percentage of additive and eliminates the hazards associated with the handling and weighing out of the individual ingredients.

Because of the dust problem which is associated with UPVC dry blends additives are now being supplied in a granular, flake or spaghetti form. Although the dispersion characteristics of the latter forms are not as good as when the additive is a powder, they are preferred by the processors due to the safety aspect.

17 Cylinder Equipment

It is recommended that the screw design be as follows. The screw is not usually fitted with a back-flow valve of the sliding ring type due to the risk of stagnation and subsequent degradation. However due to the advantages found with the back flow valve (i.e. an improvement of cycle reproducibility, the reduction in the back flow of the melt and the requirement to produce components having a good surface definition) modified designs of the sliding ring type have been developed to suit UPVC. A typical screw design has an L:D ratio of 18 to 24:1; shorter L:D ratios of 16:1 are successfully used, but for dry blend the higher ratio is preferred. Barrier design screws are also used when processing PVC dry blends to provide better homogeneity of melt. When using these types of screws the amount of heat developed can often be excessive and therefore it is usual to provide cooling within the screw particularly in the tip region. The longest section of the screw is the compression zone (usually 0.4L for a typical screw) with the feed and metering zones being 0.3L and 0.3L resp. The compression ratio ranges from 1.5 to 2.3:1 depending upon the type of compound and size of molding machine used. When using a general purpose designed screw to process UPVC, if a temperature override of 10 to 15°C (18 to 27°F) on the barrel zones 2, 3 and 4 is apparent, then the amount of shear being applied by the screw is too high. Such overrides can be reduced by either lowering the screw rotational speeds to achieve a screw surface speed of 0.15m/sec or by using a screw having a lower compression ratio. The latter choice enables a higher screw rotational speed than previous to be used, thus providing the opportunity to reduce the cycle time.

Some molders have found that lower compression ratios screws are more suitable for the larger molding machines (i.e. >500 tonnes clamp capacity) whereas the higher compression ratio screw is preferred on the smaller clamp capacity machines. The depth of flight in the metering section is important so as to prevent excessive shear being imparted to the molten material; typical values are given below.

The design of the screw tip is such that the inclusive angle

Flight depth		Diameter of screw	
mm	(in)	mm	(in)
2	(0.078)	35	(1.38)
3	(0.118)	50	(1.97)
4	(0.157)	60	(2.36)
5	(0.097)	75	(2.95)
6	(0.236)	100	(3.94)

is usually between 25 to 30 degrees and that the surface of the tip should be highly polished or of the spiral self cleaning type. The optimum clearance between the screw tip and and nozzle, when the screw is in it's full forward position, should be 0.7 to 1.8mm (0.028 to 0.072in). This small gap ensures that a only a very small amount of PVC remains in the nozzle after each injection cycle. The components of the barrel assembly which are in direct contact with hot PVC must be able to resist the corrosive nature of hydrochloric acid. To protect against such corrosion the screw, barrel and back flow valve are often produced from stainless steels containing a high chromium content. Chrome plating in conjunction with nitriding is often used for bi-metallic barrels and fully hardened screws produced from other types of steels. The cylinder assembly should be designed so that all sharp corners, abrupt changes in cross-section and direction are avoided so as to prevent the material from becoming trapped in a nil flow area causing it to degrade.

An open nozzle having a straight and perfectly smooth bore should be used. A reverse taper geometry with an entrance diameter of at least 5mm (0.197in) is recommended: keep the length of the nozzle as short as possible i.e. 51 to 102mm (2 to 4in). The exit diameter of the nozzle should be smaller than the entrance diameter of the sprue bush by 1mm (0.039in). Keep the orifice land, which includes the reverse taper, at the front of the nozzle to a minimum i.e 4 to 6mm (0.157 to 0.236in). Accurate temperature control of the nozzle is vital to avoid thermal degradation. 'Shut off' nozzles are not recommended for use with this material due to the problem of degradation. Decompression or suck back may be used however it is not recommended so as to prevent oxidation occurring to the material.

A correctly designed hopper is essential when processing dry blends: the angle of the hopper walls should be between 45 to 60 degrees. Angles of less than 45 degrees can cause bridging problems.

18 Screw Cushions

On smaller machines this should be 2 to 3mm (0.078 to 0.118in); on larger machines 9mm (0.354in) is suggested.

19 Shot Capacity

Shot capacities of 75% and above are recommended; if the screw has been designed for UPVC then up to 95% capacity can be used. The smaller the shot size, compared to the total capacity of the barrel, the longer is the material residence time and therefore the greater the risk of thermally degrading the material.

20 Melt Temperature

This is as measured in the nozzle or by an air shot technique; 185 °C to 205°C (363°F to 401°F). The correct melt temperature of UPVC can be readily identified by the glossy nature of the purgings. If the purgings have an appearance of a bumpy or lumpy nature then the material is non-homogeneous (i.e. not sufficiently plasticized) indicating that the set temperatures are too low. Whereas if the purgings have a frothy appearance and give off an excess amount of smoke then the material is too hot.

21 Barrel Residence Time

At a temperature of 200°C (392°F) the maximum residence time in the barrel should not exceed 5 minutes. When the temperature is 210°C (419°F) then the maximum residence time in the barrel should not exceed 3 minutes.

22 Temperature Settings

Please note that it is the melt temperature which is important; those temperatures given in the table are only suggested, initial settings. The temperature of the hydraulic oil, the hopper block and the material temperature should also not vary excessively. Mold temperature control is extremely important and therefore these temperatures must be checked regularly. The temperature controllers should be initially set towards the lower end of the temperature settings range and then gradually be increased to achieve the desired melt temperature.

Zone No.	Location	Temperature in °C FROM	TO	Temperature in °F FROM	TO
0	Hopper throat	40	60	104	140
1	Barrel rear	140	160	284	320
2	Barrel middle	150	170	302	338
3	Barrel middle	160	180	320	356
4	Barrel front	170	190	338	374
5	Nozzle	190	215	374	419
6	Mold	30	60	86	140

Please note that it is the melt temperature which is important; those temperatures given in the table are only suggested, initial settings. The temperature of the hydraulic oil, the hopper block and the material temperature should also not vary excessively. Mold temperature control is extremely important and therefore these temperatures must be checked regularly.

23 Injection Speed

The mold fillings speeds used with UPVC need to be relatively slow otherwise excessive shear can cause the material to degrade. Profiled mold filling speeds are often used to produce thick sectioned components from UPVC which require a high gloss. Moldings which contain faint brown streaks radiating from the gate indicate that the mold filling speed used is too fast. Mold filling speeds can also be dictated to by the size of sprue bush entry. Hence the larger the diameter the faster the speed.

24 Injection Pressure

The machine should be capable of giving up to:

first stage, up to 1500bar; 150MNm^{-2}; 21,750psi.
second stage, up to 1000bar; 100MNm^{-2}; 14,500psi.

Do not overpack the material with an excessive hold pressure, otherwise the molded component may suffer from warpage or surface cracking.

25 Screw Rotational Speed

Should be adjusted to suit the molding cycle, in general, do not use a screw surface speed of greater than 0.15 to 0.2 m.s^{-1} (0.49 to 0.66ft/sec). If a screw of 40mm (1.57in) diameter is being used then the rpm should not exceed 95. Similarly, the maximum diameter of a screw rotating at 200 rpm is 19mm (0.75in) if a screw surface speed of 0.15 to 0.2 ms^{-1} (0.49 to 0.66ft/sec) is not to be exceeded. However, for screws which have been specifically designed for UPVC then screw surface speeds of up to 0.25 to 0.27ms^{-1} (0.82 to 0.89ft/sec) may be used.

26 Back Pressure

Up to 150bar; 15MNm^{-2}; 2175psi should be available.
Back pressure should be kept as low as possible: usually a nominal value of 5 bar (0.5MNm^{-2}, 72.5psi) is used so as to achieve consistent shot weights. May need to be increased, to assist dispersion, when using dry blends or colorants but keep as low as possible.

27 Shutting Down

As the thermal stability of UPVC is poor then the shutting down procedure is very important. Ensure that all traces of the UPVC have been removed from the barrel by purging thoroughly with another thermoplastic material, that is more thermally stable and does not react with UPVC (for example PMMA, PP, LDPE, HDPE or GPPS). Usual procedure for shutting down is to switch off the barrel heaters and purge the barrel clean.

NEVER MIX POLYACETAL (POM) and UPVC TOGETHER IN THE SAME BARREL as the resultant vigorous chemical reaction can either cause considerable damage to the molding machine or serious injury to molding shop personnel. If UPVC is to follow POM (or visa versa) then thoroughly purge the barrel clean with one of the materials stated above first. It is common practice to keep a separate screw and barrel assembly solely for UPVC.

28 Reprocessing

When this material is reclaimed it is suggested that up to 20% be blended with the virgin material. The end usage of the molded component dictates the level of regrind that can be used. What must be borne in mind is that each time UPVC is reprocessed it's resistance to thermal degradation is considerably reduced. Instances have occurred where moldings fail whilst in service due to the abuse of regrind usage. Problems such as these occur not only because of the percentage of regrind being incorporated but also as a result of using regrind which has been degraded as a consequence of being reprocessed too many times.

29 Finishing

UPVC can be machined without difficulty, but keep the surfaces being cut well cooled. It may be welded by various techniques such as ultrasonic, heated tool, hot gas, friction and solvent. UPVC can be decorated using various methods such as painting, hot foiling, vacuum metallizing, EMI shielding and silkscreening.

30 Other Comments

Vinyl chloride (VC) polymers may be made by emulsion (giving PVC-E), suspension (PVC-S) and bulk, also known as mass, polymerization (PVC-M). As VC is toxic the level of monomer left in the plastic is kept very low ,for example, less than 1 part per million (<1ppm). PPVC is usually based on PVC-S or, PVC-M although PVC-E plastics can be used if the particles absorb plasticizer, for example, by agglomerating the primary PVC particles into cenospheres.

Because of the developments which have taken place in

UPVC technology, for example, alloying, blending etc., the material can provide a cost effective solution for applications which formerly demanded the use of the more expensive engineering type materials.

31 Typical Components

PVC is widely used in packaging: such packaging is often disposable. A common way of disposing of waste/rubbish is to burn it in incinerators. PVC has for many years been under suspicion as being potentially harmful to health, and to the environment, as being a precursor in the formation of dioxins and furans. Incineration studies do not support this view and some people therefore believe that incineration is a safe and viable disposal option. Operating conditions during burning, rather than the material being burnt, appear to have a more significant effect on toxic gas emissions. Do not operate incinerators at temperatures below 750°C (1400°F) as otherwise high dioxin levels result. To burn any plastic does, however, seem silly and particularly so for PVC as, HCL is formed. Waste PVC can be recycled and the trend towards recycling such materials is growing.

Pipe fittings for rainwater and irrigation projects are a major outlet for injection moldings. Other products consist of housings for computers and televisions, water filtration pressure tanks, photocopier doors, electricity and gas meter housings, transparent printer hoods, ventilation grilles, and various components associated with the electronics industry. The use of UPVC alloys for computer housings has become very popular due to it's inherent flame retardant properties, excellent physical properties, U.V. stability, low price and the ability to produce large flat area components to close fitting tolerances.

32 Unit Price

If the price of a natural, unmodified grade of GPPS is taken as one then the relative price of this molding material is approximately:

UPVC grade for Pipe fittings:	1.13
UPVC grade for High performance:	1.55

The relative price compared to GPPS can also be found in Table 7.

TABLE 1. SOME NAMES AND ABBREVIATIONS OF PLASTICS AND ELASTOMERS.

Common name	Abbreviation
Acetal (homopolymer and copolymer)	POM-H and POM-K
Acrylate styrene acrylonitrile	ASA or AAS
Acrylate modified styrene acrylonitrile	ASA or AAS
Acrylic acid ester rubber	ACM
Acrylonitrile butadiene rubber or nitrile butadiene rubber	NBR
Acrylonitrile butadiene styrene	ABS
Acrylonitrile styrene/chlorinated polyethylene	ACS
Acrylonitrile methyl methacrylate	AMMA
Acrylonitrile styrene/EPR rubber or, acrylonitrile ethylene propylene styrene	AES
Alpha methyl styrene	AMS
Atactic polypropylene	APP or PP-A
Butadiene rubber or, cis-1,4-polybutadiene rubber or, polybutadiene rubber	BR
Butadiene styrene block copolymer	BDS
Butyl rubber	IIR
Bulk molding compound	BMC
Casein formaldehyde	CF
Cellulose acetate	CA
Cellulose acetate butyrate	CAB
Cellulose acetate propionate	CAP
Cellulose nitrate	CN
Chlorinated polyethylene	CPE or CM
Chlorinated polyvinyl chloride	CPVC or, PVC-C
Chloro-polyethylene or, chlorinated polyethylene.	CM or CPE or, PE-C
Chloroprene rubber or, polychloroprene rubber	CR
Chlorotrifluoroethylene ethylene copolymers	ECTFE
Cis-polyisoprene or, cis-1,4-polyisoprene	IR
Coumarone indene resins	CIR
Diallyl phthalate	DAP
Diallyl isophthalate	DAIP
Dough molding compound	DMC
Elastomeric alloy melt processable rubber	EA-MPR
Elastomeric alloy thermoplastic vulcanizate	EA-TPV
Epichlohydrin rubber	CHR
Epoxy or, epoxide	EP
Epoxy or, epoxide, with glass fiber	EP GF
Ethyl cellulose	EC
Ethylene acryic acid	EAA
Ethylene propylene diene monomer (an EPR terpolymer)	EPDM
Ethylene propylene monomer (an EPR copolymer)	EPM
Ethylene propylene rubber	EPR or EP(D)M
Ethylene tetrafluoroethylene copolymer)	ETFE

Ethylene vinyl acetate	EVA
Ethylene vinyl alcohol	EVOH or EVAL
Expanded polystyrene	EPS or PS-E or XP or PS-X
Fiber reinforced epoxy or epoxide	FRE
Fiber reinforced plastic	FRP
Fluorinated ethylene propylene copolymer	FEP or TFE-HFP
Glass mat reinforced plastic	GMT
Granular polyester molding compound	GPMC
High density polyethylene	HDPE or PE-HD
High impact polystyrene	HIPS or TPS or IPS
Isobutene-isoprene rubber	IIR
Linear low density polyethylene	LLDPE or PE-LLD
Liquid crystal polymer	LCP
Low density polyethylene	LDPE or PE-LD
Medium density polyethylene	MDPE or PE-MD
Melamine formaldehyde	MF
Melamine phenol formaldehyde	MPF
Melt processable rubber	MPR or EA-MPR
Methyl methacrylate, styrene/polybutadiene	MBS
Natural rubber	NR
Nitrile butadiene rubber or, acrylonitrile butadiene rubber	NBR
Olefin thermoplastic elastomer	TPO
Oriented polyethylene terephthalate	OPET
Oriented polypropylene	OPP
Oriented polystyrene	OPS
Oriented polyvinyl chloride	OPVC
Perfluoroalkoxy copolymers	PFA
Phenol formaldehyde	PF
Phenylene ether copolymer	PEC
Plasticized polyvinyl chloride	PPVC or PVC-P
Polyacrylonitrile	PAN
Polyamide	PA
Polyamide 6 or nylon 6	PA6
Polyamide 11 or nylon 11	PA11
Polyamide 12 or nylon 12	PA12
Polyamide 66 or nylon 66	PA66
Polyamide 610 or nylon 610	PA610
Polyamide-imides	PAI
Polyaryl amide	PAA 6 or PAMXD6
Polybenzimidazole	PBI
Polybutadiene rubber	BR
Polybutylene	PB
Polybutylene terephthalate	PBT
Polycarbonate	PC
Polychloroprene rubber or, chloroprene rubber	CR
Polychlorotrifluorethylene	PCTFE
Polyether block amide	PEBA
Polyether ether ketone	PEEK
Polyether ester	PEEL or COPE or YPBO or TEEE

Polyether imide	PEI
Polyether ketone	PEK
Polyether ketone ketone	PEKK
Polyether sulfone or polysulfone	PES or PSU
Polyethylene	PE
Polyethylene-high density	HDPE or PE-HD
Polyethylene-linear low density	LLDPE or PE-LLD
Polyethylene-low density	LDPE or PE-LD
Polyethylene-medium density	VLDPE or PE-VLD
Polyethylene terephthalate	PET
Polyethylene terephthalate glycol	PETG
Polymethyl methacrylate (acrylic)	PMMA
Polynorbornene (rubber)	PNR
Polyimides	PI
Polyoxymethylene or, acetal or, polyformaldehyde POM	
Polyphenylene ether	PPE
Polyphenylene oxide (modified)	PPO or PPO-M or PPE
Polyphenylene sulfide	PPS
Polyphenylene sulfide sulfone	PPPS
Polyphenylene sulfone	PPSU
Polypropylene	PP or PPR
Polypropylene copolymer	PP-K
Polypropylene homopolymer	PP-H
Polypropylene oxide	PPOX
Polystyrene	PS or GPPS
Polytetrafluorethylene	PTFE
Polyurethane	PUR
Polyvinyl acetate	PVA or PVAC
Polyvinyl alcohol	PVAL
Polyvinyl carbazole	PVCZ
Polyvinyl chloride	PVC
Polyvinyl chloride acetate copolymer	PVCA
Polyvinyl pyrolidone	PVP
Polyvinylidene chloride copolymers	PVDC
Polyvinylidene fluoride	PVDF
Polyvinyl fluoride	PVF
Rubber modified polypropylene	RRPP or, RMPP or, PP/EPDM
Rubber reinforced polypropylene	RRPP or, RMPP or, PP/EPDM
Rubber reinforced polypropylene– with crosslinked rubber	TPO-XL (an EA-TPV or TPV)
Silicone plastics	SI
Styrene acrylonitrile copolymer	SAN
Styrene-butadiene rubber	SBR or, GRS
Styrene butadiene styrene block copolymer	SBS
Sheet molding compound	SMC
Styrene acrylonitrile copolymer	SAN
Styrene butadiene block copolymer	BDS or SBB

Styrene butadiene styrene block copolymer – a thermoplastic elastomer	SBS
Styrene butadiene styrene block copolymer – a saturated thermoplastic elastomer	SEBS
Styrene maleic anhydride	SMA
Synthetic polyisoprene	IR
Tetrafluoroethylene-ethylene copolymers	ETFE
Thick molding compound.	TMC
Thermoplastic elastomer (or rubber)	TPE or TPR
Thermoplastic elastomer - amide based	TPE-A
Thermoplastic elastomer - olefin based	TPE-O
Thermoplastic elastomer - urethane based	TPE-U
Thermoplastic ether ester (COPE or PEEL)	TP-EE
Thermoplastic ethylene propylene rubber – a TPE	TP-EPDM
Thermoplastic ethylene vinyl acetate elastomer – a TPE	TP-EVA
Thermoplastic natural rubber - a TPE	TP-NR
Thermoplastic nitrile butadiene rubber	TP-NBR
Thermoplastic polyolefin	TPO
Thermoplastic polyolefin- with crosslinked rubber	TPO-XL
Thermoplastic polyurethane	TPU
Thermoplastic rubber (or elastomer)	TPR or TPE
Thermoplastic vulcanizate (a TPE)	TPV or EA-TPV
Unplasticized polyvinyl chloride	UPVC or PVC-U
Unsaturated polyester	UP
Unsaturated polyester, with glass fiber	UP GF (PMC, DMC or SMC)
Urea formaldehyde	UF
Very low density polyethylene	VLDPE or PE-VLD
Vinyl chloride ethylene	VCE
Vinyl chloride ethylene vinyl acetae	VCEVA
Vinyl chloride vinyl acetae	VCVA
Vinyl ester resins	VE
Vinylidene chloride, acrylonitrile copolymers	VCA

TABLE 2A. LETTERS USED TO MODIFY ABBREVIATIONS FOR PLASTICS (ISO AND ASTM).

Letter	Meaning or significance
C	Chlorinated
D	Density
E	Expandable or expanded
F	Flexible or fluid/liquid state
H	High
I	Impact
L	Linear or low
M	Medium or molecular
N	Normal or novolak
P	Plasticized or plasticised
R	Raised or resol
T	Thermoplastic
U	Ultra or, unplasticized or unplasticised
V	Very
W	Weight
X	Crosslinked or crosslinkable

TABLE 2B. COMMONLY-USED LETTERS USED TO MODIFY ABBREVIATIONS FOR PLASTICS (I.E. IN ADDITION TO TABLE 2A).

Letter	Meaning or significance
A	Atactic or, amorphous
B	Block copolymer
C	Crystalline
E	Emulsion (polymer)
F	Fiber/fibre
G	Glass
H	Homopolymer
K	Copolymer
M	Mass or bulk (polymer) or, mat
O	Oriented
R	Random copolymer or, reinforced
S	Suspension (polymerization)
V	Vulcanized or crosslinked
EP	Engineering thermoplastic or, engineering thermoplastics material
FR	Flame retardant
GF	Glass fiber
GMT	Glass mat (reinforced) thermoplastics (material)
GP	General purpose
HI	High impact
PMC	Polyester molding compound
TP	Thermoplastic
XL	Crosslinked or, cured or, vulcanized

TABLE 2C. SYMBOLS USED FOR FILLERS AND/OR REINFORCING MATERIALS.

Letter	Meaning or significance
A	Asbestos
B	Boron or, beads or, spheres or, balls
C	Carbon or, chips, or, cuttings
D	Powder
F	Fiber/fibre
G	Glass or, ground
H	Whisker
K	Chalk or, knitted fabric
L	Cellulose or, layer
M	Mineral or, metal
N	Non-woven fabric (usually thin)
P	Mica or, paper
Q	Silicon
R	Aramid or, roving
S	Synthetic, organic or, scale, flake
T	Talcum or, cord
V	Veneer
W	Wood
Y	Yarn
Z	Others.

Please note that if a letter is not being used in these tables then, it does not mean that it is not being used in another branch of the plastics industry.

TABLE 3. SOME ABBREVIATIONS AND TRADE NAMES OF PLASTICS.

Abbreviation	Common name	Common trade names or, trade marks
ABS	Acrylonitrile butadiene styrene	Cycolac; Lustran
AMS	Alpha methyl styrene	Elite HH
ASA	Acrylonitile styrene acrylonitrile (AAS)	Luran S
BDS	Butadiene styrene block copolymer	K resin; Styrolux
BMC	Bulk molding compound	Freemix; Norsomi
CA	Cellulose acetate	Cellidor; Tenite
CAB	Cellulose acetate butyrate	Cellidor; Tenite
CAP	Cellulose acetate propionate	Cellidor; Tenite
CF	Casein formaldehyde	Erinoid; Lactoi
CN	Cellulose nitrate	Celluloid; Xylonite
COPE	Polyether ester elastomer	Arnitel; Hytrel
CP	Cellulose propionate (CAP)	Cellidor; Tenite
CPE	Chlorinated polyethylene (PE-C)	Bayer CM; Tyrin CM
CPVC	Chlorinated polyvinyl chloride	(PVC-C)
DAP	Diallylphthalate	
DAIP	Diallyl isophthalate	
DMC	Dough molding compound	Beetle DMC; ERF DMC
EA-MPR	Elastomer alloy melt processable rubber	Alcryn
EA-TPV	Elastomer alloy thermoplastic vulcanizate	Lomod; Santoprene
EP	Epoxide or epoxy (cured)	Araldite
ETFE	Tetrafluorethylene-ethylene copolymers	Tefzel
EVA	Ethylene vinyl acetate copolymer (EVAC)	Evatane
EVAL	Ethylene vinyl alcohol copolymer	Clarene; Eval
EVOH	Ethylene vinyl alcohol copolymer	Clarene; Eval
FEP	Fluorinated ethylene propylene (TFE-HFP)	Teflon FEP
GPMC	Granular polyester molding compound	Freeflo; Impel
HDPE	High density polyethylene (PE-HD)	Lupolen HD; Rigidex
HIPS	High impact polystyrene (TPS or IPS)	Lustrex; Polystyrol
LCP	Liquid crystal polymer	Vectra; Xydar.
LDPE	Low density polyethylene (PE-LD)	Alathon; Hostalen LD
MDPE	Medium density polyethylene (PE-MD)	Fortiflex
MF	Melamine formaldehyde	Melmex; Melopas
MPR	Melt processable rubber	Alcryn
PA	Polyamide or nylon.	
PA 6	Polyamide 6 or nylon 6	Akulon K; Ultramid
PA 11	Polyamide 11 or nylon 11	Rilsan B
PA 12	Polyamide 12 or nylon 12	Rilsan A; Grilamid
PA 46	Polyamide 46 or nylon 46	Stanyl
PA 66	Polyamide 66 or nylon 66	Maranyl; Zytel
PA 610	Polyamide 610 or nylon 610	Brulon; Perlon
PAA 6	Polyaryl amide or, poly-m-xylene-adipamide (PA MXD6)	Ixef

Abbreviation	Common name	Common trade names or, trade marks
PAN	Polyacrylonitrile	Acrilan; Barex; Orlon
PBI	Polybenzimidazole	Celazole
PBT	Polybutylene terephthalate	Pocan; Valox
PC	Polycarbonate	Lexan; Makrolon
PCTFE	Polychlorotrifluorethylene	Hostaflon C2; Kel-F
PE	Polyethylene	Alathon; Lupolen
PEBA	Polyether block amide	Pebax
PEEK	Polyether ether ketone	Victrex PEEK;
PEEL	Polyether ester (YPBO)	Arnitel; Hytrel
PE-HD	Polyethylene-high density	Lupolen HD; Rigidex HDPE
PEI	Polyether imide	Ultem.
PEK	Polyether ketone	Hostatec
PEKK	Polyether ketone ketone	(from DuPont)
PE-LD	Polyethylene- low density	Alathon; Lupolen
PE-MD	Polyethylene- medium density	Fortiflex
PE-VLD	Polyethylene- very low density	Norsoflex
PET	Polyethylene terephthalate	Arnite A; Techster E
PES	Polyether sulphone	Victrex.
PMC	Polyester molding compound	Aropol; Norsomix
PF	Phenol formaldehyde	Bakelite; Sternite
PI	Polyimide	Vespel
PMMA	Polymethyl methacrylate (acrylic)	Diakon; Plexiglas
PMMA-T	Toughened acrylic	
POM	Polyoxymethylene or, acetal or, polyformaldehyde	Delrin; Hostaform
POM-H	Acetal homopolymer	Delrin and Delrin II
POM-K	Acetal copolymer	Hostaform; Ultraform
PP	Polypropylene	Profax; Propathene
PPE	Polyphenylene ether (see PPO)	
PPO	Polyphenylene oxide – usually modified polyphenylene oxide (PPO-M)	Luranyl; Noryl
PPS	Polyphenylene sulphide	Fortron; Ryton
PPPS	Polyphenylene sulphide sulphone	Ryton S
PPVC	Plasticised polyvinyl chloride (PVC-P)	Solvic; Vinnol
PS	Polystyrene (GPPS)	Lustrex; Polystyrol
PSU	Polysulphone	Udel
PTFE	Polytetrafluorethylene	Fluon; Teflon
PVC	Polyvinyl chloride	Corvic; Geon
PVDC	Polyvinylidene chloride copolymers	Saran
PVDF	Polyvinylidene fluoride	Dyflor; Kynar; Solef
PVF	Polyvinyl fluoride	Tedlar
SAN	Styrene acrylonitrile copolymer	Lustran SAN; Tyril

Abbreviation	Common name	Common trade names or, trade marks
SMC	Sheet molding compound	ERF SMC; Flomat
UPVC	Unplasticised polyvinyl chloride (PVC-U)	Corvic; Geon
RMPP	Rubber modified polypropylene (PP/EPDM)	Uniroyal TPR; Keltan
RRPP	Rubber reinforced polypropylene (PP/EPDM)	Uniroyal TPR; Keltan
SBS	Styrene butadiene styrene block copolymer	Cariflex TR; Solprene
SEBS	Styrene butadiene styrene block copolymer (saturated)	Cariflex
TPE	Thermoplastic elastomer (rubber)	
TP-EE	Thermoplastic elastomer – ether ester	Arnitel; Hytrel
TP-EPDM	Thermoplastic elastomer – based on EPDM	
TP-EVA	Thermoplastic elastomer – based on EVA	
TP-NBR	Thermoplastic elastomer - based on NBR	
TPO	Thermoplastic polyolefin	Propathene OTE; Vistaflex
TPO-XL	Thermoplastic polyolefin rubber – crosslinked (rubber)	Levaflex; Santoprene
TPR	Thermoplastic rubber (elastomer)	
TPU	Thermoplastic polyurethane	Elastollan; Estane
TPV	Thermoplastic elastomer or rubber – crosslinked (rubber)	Lomod; Santoprene
UF	Urea formaldehyde	Beetle; Scarab
VE	Vinyl ester resins	
VLDPE	Very low density polyethylene	Norsoflex.

TABLE 4 SOME TRADE NAMES, ABBREVIATIONS AND SUPPLIERS OF POLYMERS AND POLYMER COMPOUNDS.

Trade name/ trade marks	Abbreviation	Supplier
A-C	Low mol wt. PE	Allied Signal Imc.
A-Clyn	Low mol wt. ionomers	Allied Signal Imc.
Acrylite	PMMA	Canada Colors & Chemicals
Adpro	PP	Advanced Global Polymers
A-Fax	PP-A	Himont
Akulon K	PA6	Akzo Engineering Plastics
Akulon M	PA6	Akzo Engineering Plastics
Alathon	PE-HD	DuPont
Alcoryl	ABS	Rhone Poulenc
Alcryn	EA-TPV or TPE	DuPont
Algoflon	PTFE	Enimont
Altulite	PMMA	Altulor
Amoco HDPE	PE-HD	Amoco
Apec	Polyester PC	Bayer
Appryl	PP	Appryl
Apscom	Thermoplastics compounds	Akzo Engineering Plastics
Araldite	EMC or EP	Ciba Geigy
Ardel	Polyarylate	Amoco
Arnite A	PET	Akzo Engineering Plastics
Arnitel	PEEL or COPE	Akzo Engineering Plastics
Arnite	PBT	Akzo Engineering Plastics
Aropol	PMC	Ashland Chemical Co.
Arylon T	ABS/PSU	USS Chemicals
Ashlene	PA 66	Ashley Polymers Inc
Azdel	GMT/PP	GEP (General Electric Plastics)
Azloy	GMT/PC/PBT	GEP (General Electric Plastics)
Azmet	GMT/PBT	GEP (General Electric Plastics)
Bakelite	PF	Bakelite/Sterling Molding Materials
Bakelite Polyester Alkyd	GPMC	Sterling Molding Materials

Trade name/trade marks	Abbreviation	Supplier
Bapolan	PS	BASF/Bamberger
Bapolan	HIPS	BASF
Bapolene	HDPE & PP	Bamberger
Barex	PAN	Sohio
Barex	PAN	Standard Oil
Bayblend	ABS/PC	Bayer
Bayer CM	CPE	Bayer
Baycomp	Fiber filled thermoplastics	Baycomp
Baygal	Encapsulating resins	Bayer
Baymidur	Encapsulating resins	Bayer
Baypren	CR	Bayer
Beetle	UF	BIPChemicals/Cyanamid
Beetle DMC	DMC	BIP Chemicals
Beetle nylon 6	PA 6	BIP Chemicals
Beetle nylon 66	PA 66	BIP Chemicals
Beetle PET	PET	BIP Chemicals
Bergamid A	PA66	Bergmann
Bergamid B	PA6	Bergmann
Bergacell	CA	Bergmann
Bexloy	Engineering thermoplastics for cars	DuPont
Buna AP	EPM/EPDM	Bunawerke Huls
Buna CB	BR	Bayer
Buna EM	SBR	Bunawerke Huls
Cabelec	PP (conductive)	Cabot Plastics
Cadon	SMA	Monsanto
Cadon 300	Impact modified SMA (used with PVC)	Monsanto
Calibre	PC	Dow
Caprolan	TPU	Elastogran/BASF
Capron	PA6	Allied Corp
Carbaicar	UF	S A Aicar
Carbopol	Acrylic acid polymer	BF Goodrich
Carboset	Acrylic polymer	BF Goodrich
Carinex	HIPS	Shell
Cariflex	TPE (SBS)	Shell
Celanese Nylon	PA 66	Hoechst/Hoechst Celanese
Celanex	PBT	Hoechst/Hoechst Celanese
Celazole	PBI	Hoechst/Hoechst Celanese
Cellidor	CA & CAP	Albis
Celsir	UF resins	SIR
Celstran	Long fiber filled thermoplastic	Hoechst/Hoechst Celanese
Cellobond BP	PF	BP Chemicals

Trade name/ trade marks	Abbreviation	Supplier
Centrex	ASA or AAS	Monsanto
Clarene	EVOH or EVAL	Solvay
Corton	PP mineral filled	Poly Pacific Pty
Corvic	PVC	EVC
Crastine	PBT	Ciba Geigy
Craston PPS	PPS	Ciba Geigy
Crystic	UP	Scott Bader
Crystic Impreg	GMC	Scott Bader
Cycolac	ABS	Monsanto
Cymel	MF	Cyanamid
DAP 5000	DAP	Synres Amoco
DAIP 6000	DAIP	Synres Amoco
Daplen	LDPE: HDPE: PP	Chemie Linz
Delrin	POM-H	DuPont
Delrin II	POM-H	DuPont
DER	EP	Chemroy Canada
Desmopan	TPU	Bayer
Dexel and Dexel S	CA	Courtaulds Speciality Plastics
Dexflex	TPO	Dexter Plastics
Diakon	PMMA	ICI
Dialac	ASA	Mitsubishi
Dion	UP	Fiberglass
Dowlex	PE-LD	Dow Chemicals
Dowlex	PE-LLD	Dow Chemicals
DSM Nyrim	Reaction injection molding (RIM) PA	DSM RIM Nylon
Duracryn	TPE/TPV	DuPont
Dural	reinforced UPVC	Dexter Plasics
Duralex	PVC/PU/NBR alloy	Dexter Plasics
Duralon	PA 11	Thermoclad
Durapol	PMC	Isola Werke
Durez	PF	Occidental/ Canadian Oxy Chemicals
Durez	DAP	Occidental
Durolon	PC	Atochem
Dutral	PS	Enimont
Dutral FLX	TPE/TPV	Enimont
Dyflor	PVDF	Dynamit Nobel
Dyflor	PVDF	Kay Fries
Dynaform	RRPP	Dynamit Nobel
Dynaset	PF	Reichold
Ecdel	COPE or PEEL	Eastman Chemicals
Eccomold	EP	Emmerson and Cuming
Ecolyte II & IV	PE	Ecoplastics/ Eco Chemicals

Trade name/ trade marks	Abbreviation	Supplier
Ecolyte	Photodegradable TP	Ecoplastics/Eco Chemicals
Ecolyte S	PS	Ecoplastics/Eco Chemicals
Edister	ABS	Enimont
Edister	HIPS	Enimont
Edister	PS	Enimont
Elastar	TPE (based on NBR-XL/PVC)	
Elastollan	TPU	Elastogran/BASF
Elastopreg	GF/GMT laminates	BASF/Elastogran
Elemid	PPO-M/PA	Borg Warner
Elite HH	AMS	Monsanto
Elvanol	EVAL	DuPont
Elvax	EVA copolymer	DSM
Empee PP	PP	Monmouth Plastics
Envex	Reinforced PI	Rogers Corp
Epoester	EP resins	SIR
Epolan	ABS	Industrial Resistol
Epolene	Maleated PP	Eastman Chemicals
Epon	EP	Shell
Eponac	EP	SPREA
Eposir	EP resins	SIR/ Kingsley & Keith
ERF DMC	DMC	ERF
ERF SMC	SMC	ER
Ertalan	cast PA	Erta
Escorene	PE-LLD	Exxon
Estane	TPU	Goodrich
Estamid	PEBA/TPE	Dow
Esterform	PMC	Chromos Ro-Polimeri
ETA	TPE	Republic Plastics
Europrene	TPE (SBS)	Ani
Eval	EVOH or EVAL	Kuraray/EVAL Co.
Exxelor	Plastics modifiers	Exxon
Fenochem	PF	Chemiplastica Spa
Fenoform	PF	Chromos Ro-Polimeri
Fibiter	PBT	Enimont
Flomat	SMC	Freeman Chemicals
Foraflon	PVDF	Atochem
Fortiflex	PE-MD	Soltex Polymer Corp.
Fortiflex	PE-HD	Soltex Polymer Corp.
Fortilene	PP	Soltex Polymer Corp.
Fortron	PPS	Hoechst/Hoechst Celanese
Freeflow	GPMC	Freeman Chemicals
Flowmat	SMC	Freeman Chemicals

Trade name/ trade marks	Abbreviation	Supplier
Gaflex	COPE or PEEL	Hoechst/Hoechst Celanese
Gedex	PS	Orkem
Gedex	HIPS/TPS	Orkem
Geloy	ASA	General Electric Plastics/GEP
Genal	PF	General Electric Plastics/GEP
Geolast	TPV or NBR-TPV	Monsanto
Geon	PVC	BF Goodrich
Glilax	PEBA/TPE	Dianippon
Grilamid	PA12	EMS-Grilon (EMS-Chemie)/Emser Industries
Grilon	PA 6	EMS-Grilon (EMS-Chemie)/Emser Industries
Grilon T	PA 66	EMS-Grilon (EMS-Chemie)/Emser Industries
Grivory	Amorphous EP	Emser Ind.
Halar	ECTFE	Ausimont
Halon	PTFE	Ausimont
Haysite	PMC	Haysite Reinforced Plastics
Hercules HPR	PE-HD of high molecular weight	Hercules
Hetron	PMC	Ashland Chemical Corp.
Hostaform	POM-K	Hoechst/Hoechst Celanese
Hostadur	PBT	Hoechst/Hoechst Celanese
Hostaflon FEP	FEP	Hoechst/Hoechst Celanese
Hostalen	PE-HD	Hoechst/Hoechst Celanese
Hostalen GUR	PE-UHMW	Hoechst/Hoechst Celanese
Hostalen PP	PP	Hoechst/Hoechst Celanese
Hostalen PP	reinforced PP	Hoechst/Hoechst Celanese
Hostapren	PE-C	Hoechst/Hoechst Celanese
Hostalit	PVC	Hoechst/Hoechst Celanese
Hostalit Z	PVC-HI	Hoechst/Hoechst

Trade name/ trade marks	Abbreviation	Supplier
Hostatec	PEK	Hoechst/Hoechst Celanese
Huntsman Polystyrene	PS	Huntsman
Hytrel	PEEL or COPE	DuPont
Idemitsu Polycarbonate	PC	Idemitsu Petro Chem.
Illandur	PMC	Dr. Illing
Impel	GPMC	Scott Bader
Impet	PET	Hoechst/Hoechst Celanese
Indopol	Polybutenes	Amoco
Innovex	PE-LLD	BP
Isomin	MF	Perstorp
Ixef	PAA 6	Laporte/Solvay
Iztavil	PVC	Polimeros De Mexico
Kadel	Polyketone	Amoco
Kamax	Acrylic imides	Rohm and Haas
Kane Ace	CPVC	Kaneka/ E W Seward Ltd.
Kane Ace B	MBS	Kaneka/ E W Seward Ltd.
Kane Ace PA	Acrylic proc. aid	Kaneka/ E W Seward Ltd.
Kane Ace XEL	Cross linked PVC	Kaneka/ E W Seward Ltd.
Kelburon	RRPP/TPE	DSM (Dutch State Mines)
Kelrinal	CM rubber	DSM (Dutch State Mines)
Keltan TP	RRPP/TPE	DSM (Dutch State Mines)
Keltan	EP(D)M	DSM (Dutch State Mines)
Kelprox	TPE	DSM (Dutch State Mines)
Keltaflex	TPE/TPV	DSM (Dutch State Mines)
Kematal	POM-C	Hoechst/Hoechst Celanese
Kerimid	PI	Rhone Poulenc
Keripol	PMC	Pheonix
K F	PVDF	Kureha
Kinel	PI	Rhone Poulenc
Kodapek PET	PET	Eastman Chemicals
Kodar PETG	PET-K or, PET-C	Eastman Chemicals

Trade name/trade marks	Abbreviation	Supplier
Kostil	SAN	Enimont
Kraton TR	TPE (SBS)	Shell
K resin	BDS	Phillips
Kynar	PVDF	Penwalt
Lacqrene	PS	Atochem
Lacqrene	HIPS	Atochem
Lacqtene HD	PE-HD	Atochem
Lacqtene HX	PE-LLD	Atochem
Ladene	PE-LLD	Sabic
Legupren	UP	Bayer
Leguval	UP resins	Bayer
Lekutherm	EP	Bayer
Levaflex	TPO-XL	Bayer
Levapren	EVA/PVC	Bayer
Lexan	PC	GEP (General Electric Plastics)
Linpac Polystyrene	PS	Linpac
Lomod	COPE or PEEL	GEP
Lotader	Co & terpolymers	Orkem (was CdF Chemie)
Lotrene	PE-LD	Orkem
Lotrex	PE-LLD	Orkem
Lucalor	CPVC	Orgavyl
Lucovyl	PVC	Atochem
Lupolen	PE-LD	BASF
Lupolen HD	PE-HD	BASF
Luranyl	PPO-M or PPE/HIPS	BASF
Luran	SAN	BASF
Luran S	ASA	BASF
Lustran	ABS	Monsanto
Lustran	SAN	Monsanto
Lustran Elite HH	AMS	Monsanto
Lustran Ultra ABS	ABS (high gloss)	Monsanto
Lustrex	PS	Monsanto
Lustrex	HIPS	Monsanto
Luran	SAN	BASF
Luran S	ASA or AAS	BASF
Magnum	ABS	Dow
Makroblend	PC/PBT	Bayer
Makrolon	PC	Bayer
Maranyl	PA 66	ICI/LNP
Marlex	PE-HD	Phillips
Melaicar	MF	S A Aicar
Melamine molding compounds	MF	Perstop Ferguson
Melmex	MF	BIP
Meloplas	MF	Ciba Geigy
Melsprea	MF	SPREA

Trade name/trade marks	Abbreviation	Supplier
Menzolit	PMC	Menzolit Werke
Merlin	PC	Mobay
Merlon	PC	Bayer
Metton	A liquid molding resin	Shell/Hercules
Miapol	UP resin	Mia Chemical
Mindel	PSU (filled)	Amoco Chemicals
Minlon	PA 66 (filled)	DuPont
Moldsite	PF	SPREA
Moplen RO	PE-HD	Enimont
Moplen SP	RRPP	Enimont
Morthane	TPU	Morton Thiokol Inc.
Neoflon	FEP	Daikin
Neonite	EMC or EP long glass fiber filled	Ciba Geigy
Neste HDPE	PE-HD	Neste
Neste LD	PE-LD	Neste
Neste LPLD	PE-LLD	Neste
Neste PS	HIPS	Neste
Nestorite	PF	Perstorp
Nike	CN	Punda Inc.
Norchem	PE-HD, PE-LLD & PP	Enron/Delong Prochem
Norlin	PE-LLD	Northern Petrochemicals
Norsomix	DMC/PMC	Orkem (CdF Chimie)
Norsoflex	VLDPE	Orkem (CdF Chimie)
Nortuff	PP	Norchem Inc
Norvinyl	PVC	Norsk Hydro
Noryl	PPO-M	GEP(General Electric Plastics)
Noryl GTX	PPO-M/PA	GEP(General Electric Plastics)
Novamid	a PA	Mitsuibishi
Novablend	PVC	Novatec
Novapol HD	PE-HD	Novacor Chemicals
Novapol LD	PE-LD	Novacor Chemicals
Novex	PE-LD	BP Chemicals
Novodur	ABS	Bayer
Nuloy	PA6	Terlon Polimeros
Nydur	PA6 & PA66	Mobay
Nylafil	Filled PA	Wilson Fibrefil
Nylon	PA	e.g. DuPont
Nypel	PA6	Allied Signal Inc.
Orgalloy R	PA6/PP or PA66/PP	Atochem
Orgamide	PA 6	Atochem
Orgater	PBT	Atochem
Oroglas	PMMA	Rohm and Haas

Trade name/trade marks	Abbreviation	Supplier
Palatal	UP	BASF
Paraloid	Impact modifiers	Rohm and Haas
Parr	DAP	US Prolam Inc.
Paxon	PE-HD	Allied
Pax-Purge	Purge compounds	Canada Color & Chemicals
Pebax	PEBA a TPE	Atochem
Pemex	LDPE	Petroleos Mexicanos
Pekema	PVC	Punda Inc
Pekevic	PVC	Neste Oy
Perbunan N	NBR	Bayer
Petlon	PET	Bayer/Mobay
Petra	PET	Allied Signal Inc.
Petrothene	PE-HD	USI/Quantum
Petrothene	PE-LLD	USI/Quantum
Pevikon	PVC	Norsk Hydro
Pibiflex	PEEL or COPE	Dutral/Montedison
Pibiter	PBT	Dutral/Montedison
Plasticlean	Purging compound	W S Wood Assoc.
Plaskon	EP	Plaskon Molding Div.
Plaskon DAP	DAP	Plaskon Molding Div.
Plenco	PF, MPF & UP	Plastics Engineering Co.
Plexiglas	PMMA	Rohm and Haas
Pocan	PBT	Bayer/Mobay
Polloplas	UF	Dynamit Nobel
Polyclear	PET	Hoechst/Hoechst Celanese
Polychem	DAP	Budd Co.
Polycol	PVC	Atochem
Poly DAP	DAP	US Prolam Inc.
Polyfort	PP filled	Schulman Inc
Polykemi	PC	Atochem
Polyloy	PA 6	Illing
Polyloy	PA 66	Illing
Polymer E	LDPE	Asia Polymer Corp
Polyset	EP	Morton Chemical Div.
Polystyrol	PS	BASF
Polystyrol	PS	Norsk Hydro
Polystyrol	HIPS	BASF
Polystyrol	HIPS	Norsk Hydro
Polystyron	PS	Svenska
Polyvest	Polybutadiene	Huls
Prevex	PPO-M or PPE	Borg Warner
Primax	PE-UHMW	Air Products and Chemicals
Procom	PP compounds	ICI
Profax	PP	Himont/Hercules
Progilite	PF	Rhone Poulenc
Proloy	EP	GEP(General Electric Plastics)

Trade name/trade marks	Abbreviation	Supplier
Propathene OTE	RRPP	ICI
Pulse	ABS/PC	Dow
Quimcel	CN	Punda Inc.
Radel	PSU	Amoco
Radlite	GMT	Azdel Europe
Ravikral	ABS	Enichem
Resarit	PMMA	Resart
Resartherm	PMC	Resart
Resilon	PVC	Canadian General-Tower
Resin 18	AMS	Amoco
Rexene	PP	El Paso
Rigidex HDPE	PE-HD	BP
Rilsan A	PA 12	Atochem
Rilsan	PA 11	Atochem
Riteflex	PEEL or COPE	Hoechst/Hoechst Celanese
Ronfalin	ABS	DSM (Dutch State Mines)
Ronfaloy	ABS alloy	DSM (Dutch State Mines)
Ronfaloy E	ABS/EP(D)M	DSM (Dutch State Mines)
Ronfaloy V	ABS/PVC	DSM (Dutch State Mines)
Rovel	Styrene based material	Dow
Royalene	EPDM	Uniroyal
Rutaform	MF	Sterling Molding Materials
Rutaform Polyester	GPMC	Sterling Molding Materials
Rutamid 6	PA6	Bakelite
Rutamid 66	PA66	Bakelite
Rynite	PET	DuPont
Ryton	PPS	Phillips Chemicals
Ryton S	PPPS	Phillips Chemicals
Saran	PVDC or PVdC	Dow
Santoprene	TPO-XL an EA-TPV	Monsanto
Scarab	UF	BIP
Scarnol	EVOH or EVAL	Nippon Gobsei
Sclair	PE-LLD	DuPont
Sclairlink	Cross linkable PE	DuPont
Selar	amorphous PA	DuPont
Sinvet	PC	Enimont
Sirester	UP resins	SIR (Societa Italiana Resine)

Trade name/trade marks	Abbreviation	Supplier
Sirfen	PF resins	SIR (Societa Italiana Resine)
Sirfen X	PF	SIR (Societa Italiana Resine)
Siritle	UF	SIR (Societa Italiana Resine)
SMA Resins	SMA	Sartomer
Siritle	UF	SIR (Societa Italiana Resine)
Solef	PVDF	Laporte
Solef	PVDF	Solvay
Sol T	SBS & SIS	Enimont
Stamylan HD	PE-HD	DSM (Dutch State Mines)
Stamylan LD	PE-LD	DSM (Dutch State Mines)
Stamylan P	PP	DSM (Dutch State Mines)
Stamylex	PE-LLD	DSM (Dutch State Mines)
Stamyroid	modified amorphous PP	DSM (Dutch State Mines)
Stanyl	PA 46	DSM (Dutch State Mines)
Stapron S	rubber modified SMA	DSM (Dutch State Mines)
Statoil	PE-HD	Statoil
Statoil polyethylene LDPE	PE-LD	Statoil
Sternite	PF	Sterling Molding Materials
Sternite	PS	Sterling Molding Materials
Sternite	HIPS	Sterling Molding Materials
Styrolux	BDS	BASF
Styron	PS	Dow
Styron	HIPS	Dow
Supec	PPS	GEP (General Electric Plastics)
Supraplas	EP	Sud West Chemie
Suramin	UF and MF resins	SIR (Societa Italiana Resine)
Surlyn	Ionomer resins	DuPont
Tancin	PP	Washington Penn Plastics
Technyl B	PA 66	Rhone Poulenc
Techster E	PET	Rhone Poulenc
Techster T	PBT	Rhone Poulenc

Trade name/trade marks	Abbreviation	Supplier
Tecolit	PF	Toshiba Chemical Products
Tedur	PPS	Bayer
Teflon FEP	FEP	DuPont
Teflex	FEP	Nitechim
Tenite	CA	Eastman Chemical
Tenite	CAB	Eastman Chemical
Tenite	CP or CAP	Eastman Chemical
Tenite polyethylene	PE-LD	Eastman Chemicals
Terblend S	ASA/PC	BASF
Terluran	ABS	BASF
Therban	NBR or HNBR	Bayer
Thermocomp	PES/PSU	LNP Plastics
Thermocomp	Filled PA	LNP Plastics
Torlon	PAI	Amoco
Toyobo MXDA	PAMXD6	Toyobo
Trefsin	TPE/TPV	Exxon
Triax	PA/ABS alloys	Monsanto
Trithene	LDPE	Petroquimica Triunfo
Trithera	EVA	Petroquimica Triunfo
Trogamid T	amorphous PA	Huls
Tyril	SAN	Dow
Tyrin	CPE	Dow
Udel	PSU	Amoco Chemicals
Ugikral	ABS	Orkem
Ultem	PEI	GEP (General Electric Plastics)
Ultrablend	PBT/PC	BASF
Ultrablend S	PBT/ASA	BASF
Ultradur	PBT	BASF
Ultraform	POM-K	BASF
Ultralen	PET	BASF
Ultraplas	MF	Dynamit Nobel
Ultramid	PA 6	BASF
Ultramid A	PA 66	BASF
Ultramid C	PA copolymers	BASF
Ultramid S	PA 610	BASF
Ultranyl	PPO-M/PA or PPE/PA	BASF
Ultrason E	PES	BASF
Ultrason S	PSU	BASF
Urochem	UF	Chemiplastica Spa
Uroplast	UF	SterlingMoulding Materials
Urtal	ABS	Enichem
Valox	PBT	GEP (General Electric Plastics)
Vandar	Thermoplastic	Hoechst/Hoechst Celanese alloys
Vectra	LCP	Hoechst/Hoechst Celanese

Trade name/trade marks	Abbreviation	Supplier
Vedril	PMMA	Enimont
Versamid	PA	
Verton	Long fiber filled thermoplastic	ICI
Vespel	PI	DuPont
Vestamid	PA 12	Huls
Vestoblend	PPE/PA	Huls
	PBT	Huls
Vestolen A	PE-HD	Huls
Vestolen P	PP	Huls
Vestolit	PVC	Huls
Vestopal	UP	Huls
Vestoplast	amorphous PO	Huls
Vestopren	EPM	Huls
Vestoran	PPO-M or PPE	Huls
Vestyron	PS	Huls
Vibrin	UP resins	Fiberglass
Viclan	PVDC/PVC	ICI
Victrex PEEK	PEEK	ICI
Victrex PES	PES	ICI
Vinoflex	PVC	BASF
Vipla	PVC	European Vinyl Corp.
Vista	PVC	Vista Chemicals
Vistalon	EPDM	Exxon
Vitalon	PA 46	Trade name used in Japan for Stanyl
Vitax	ASA	Hitachi Chemicals
Vydox	PTFE	DuPont
Vydyne	PA	Monsanto
Vydyne R	PA 66 (reinforced)	Monsanto
Vynite	PVC/NBR	Alpha Chemical & Plastics
Vythene	PVC/PU	Alpha Chemical & Plastics
Wacker Chemie	PE-HD	Wacker Polyathylen
Welite	PBT	Wellman Inc.
Wellamid	PA6 & PA66	Wellman Inc.
Welpet	PET	Wellman Inc.
Xantar	PC	DSM (Dutch State Mines)
Xenoy	PC/PBT	GEP (General Electric Plastics)
Xydar	LCP	Amoco
Zytel	PA 66	DuPont
Zytel ST		PA 66 super tough DuPont

TABLE 5. WATER ABSORPTION AND DRYING CONDITIONS.

Full name	Abb.	Water abbs.		Hot air drying			Dessicant dry	
		%	°C	°F	No. of h	°C	°F	No. of h
Ethylene vinyl acetate	EVA	0.2	50-60*	122-140	8*	50-60*	4	122-140*
High density polyethylene	HDPE	0.01	65	176	3	80	176	1-1½
Low density polyethylene	LDPE	0.02	65	149	3	80	176	1-1½
Linear low density polyethylene	LLDPE	0.02	65	149	3	85	185	1-1½
Polypropylene	PP	0.02	80	176	2-3	80	176	1-1½
Rubber modified polypropylene	PP-EPDM	0.2	80	176	2-3	90		1-1½
Plasticized polyvinyl chloride	PPVC	0.02	65	149	2	70	158	1-1½
Unplastized polyvinyl chloride	UPVC	0.02	65	149	3	80	176	1-1½

* Drying temperatures and times are dependant upon the amount of vinyl acetate which is present in the material.

TABLE 6. HEAT CONTENTS OF SOME MOLDING MATERIALS.

Abbreviation	Specific heat J/kgK	Rec.temps. Melt degC	Rec.temps. Mold. degC	Temp diff degC	Heat content J/g	Amount of water to cool 1g. c.c
EVA	3130	200	25	175	563	34
HDPE	3640	240	20	220	800	48
LDPE	3180	210	30	180	572	34
LLDPE	3206	210	40	170	545	32
PP	2930	240	50	190	557	33
PP-EPDM	2264	240	50	190	430	26
PPVC	1564	200	40	160	250	15
UPVC	1820	195	40	155	291	17

Notes:
1. The amount of heat energy per gram of thermoplastics material can be calculated using the following formula:

$$\frac{\text{specific heat (J/kgK)} \times \text{temp diff (degC)}}{1000}$$

(division by 1000 is necessary as the units of specific heat are in Joules per kilogram).

2. To calculate the amount of fluid required to cool one gram of the thermoplastics material from it's melt temperature to the mold temperature, it is necessary to equate the heat input in J/g to the heat removed by the cooling medium in c.c's.

The heat removed by the cooling medium is related to the specific heat, the maximum tolerable temperature change and the flow rate of the cooling medium.

For heat removal purposes the maximum change in temperature is 4°C (7.2°F).

Hence the heat input (J/g) = coolant flow rate (q) x specific heat x temp change where the flow rate (q) =

$$\frac{\text{heat input (J/g)}}{\text{specific heat (KJ/Kg°C)} \times \text{temp change (i.e. 4°C)}}$$

The specific heat for water is 4.186(KJ/Kg°C).

The following worked example shows how to calculate the amount of water to cool 1g of HDPE: where = q x 4.186 x 4

$$q = \frac{800}{4.186 \times 4}$$

q = 47.77cc.

The above value does not include the cycle time at which the mold is to be run at. It must be noted the coolant flow rate in c.c.'s or gallons per minute is determined by the total shot weight of the mold and the cycle time needed to produce moldings to the specified standard of quality.

The calculated value of q should be regarded as the minimum flow rate of cooling medium needed to remove the heat given off by the molten material.

TABLE 7. APPROXIMATE PRICE OF MATERIALS RELATIVE TO A GENERAL PURPOSE GRADE OF PS (GPPS = 1.00).

Full name	Abb.	Natural Compound	
Ethylene vinyl acetate	EVA	GP grade (5% VA content)	0.82
		GP grade (28% VA content)	1.13
High density polyethylene	HDPE	Easy flow grade	0.82
		Medium flow grade	0.82
		GP grade	0.82
		Stiff flow grade	0.86
Low density polyethylene	LDPE	Easy flow grade	0.82
		Medium flow grade	0.82
		Stiff flow grade	0.84
Linear low density polyethylene	LLDPE	GP Grade	0.98
Polypropylene	PP	Homopolymer	
		GP grade	0.81
		Easy flow grade	0.83
		Copolymer	
		GP grade	0.87
Rubber modified polypropylene	PP-EPDM	Automotive bumper grade	1.36
Plasticized polyvinyl chloride	PPVC	GP grade	0.8 to 0.9
		Medical grade	1.2 to 1.5
Unplasticized polyvinyl chloride	UPVC	Pipe fitting grade	1.13
		High performance grade	1.55

TABLE 8. DENSITIES AND APPROXIMATE COSTS OF MATERIALS.

Abbreviation	Specific gravity	Specific volume		Approximate cost	
		cc/g	oz/cu in	p/g	c/cu in
EVA	0.926 to 0.95	1.05 to 1.08	0.54 to 0.55	0.72 to 0.99	17.48 to 24.66
HDPE	0.94 to 0.965	1.04 to 1.06	0.54 to 0.56	0.72 to 0.76	17.74 to 19.23
LDPE	0.92	1.09	0.53	0.72 to 0.74	17.36 to 17.85
LLDPE	0.90 to 0.94	1.06 to 1.11	0.52 to 0.54	0.86	20.29 to 21.19
PP	0.90	1.11	0.52	0.71 to 0.77	16.75 to 18.17
PP-EPDM	0.86 to 0.90	1.11 to 1.16	0.50 to 0.52	1.20	27.05 to 28.31
PPVC	1.25 to 1.65	0.61 to 0.80	0.72 to 0.95	0.70 to 1.32	22.94 to 57.11
UPVC	1.34 to 1.45	0.69 to 0.75	0.77 to 0.84	0.99 to 1.36	34.78 to 51.70

As plastics materials are bought by weight and sold by volume (in the form of moldings) then, any material with a low density, or specific gravity (SG), has a tremendous advantage. Often therefore, the cost per unit volume is more important than the cost per unit weight. To convert to the cost in cents per cubic inch (c/cu in), multiply the cost per pound (lb) in cents by the SG and then, by 0.0361. Note that, the density in g/cc has the same numerical value as the SG.

To obtain density in oz/cu in. multiply g/cc by 0.5781.

To obtain density in lbs/cu in. multiply g/cc by 0.0361.

The conversions from one currency to another were obtained by assuming that there are 1.60 dollars to the pound sterling.

That is, $1.60 = £1.00 or, 160 cents = 100 pence (p) or, 1.60c = 1p.

So, for example, to convert from p/g to c/g multiply the p/g by the currency conversion factor, for example, 1.60 in this case.

To convert from p/g to c/cu.in: first convert to p/lb by multiplying by 454 (454g = 1lb), then to obtain c/lb multiply p/lb by, the currency conversion factor, for example, 1.60 in this case. To convert the cost in cents per cubic inch (c/cu in), multiply the cost per pound (lb) in cents (c) by the SG and then, by 0.0361. That is:

p/g x 454 x 1.60 x 0.0361 x SG = c/cu.in.

p/g x 26.22 x SG = c/cu.in.

TABLE 9A. MATERIAL PROPERTIES.

	Property	Units	ASTM	LDPE	LLDPE	HDPE
1	Tensile strength	psi	D638	1,200-4,400	1,800-4,000	3,000-4,400
2	Elongation at break	%	D638	90-600	90-850	10-1,000
3	Tensile yield strength	psi	D638	1,200-2,000	1,400-2,500	3,500-4,700
4	Compressive strength	psi	D695	—	—	2,500-3,500
5	Flexural strength	psi	D790	—	—	—
6	Tensile modulus	10^3 psi	D638	20-40	38-75	150-155
7	Compressive modulus	10^3 psi	D695		—	—
8	Flexural modulus	10^3 psi	D790	30-50	40-70	145-225
9	Izod impact strength	'a'	D256A	No break	1-9	0.2-4.0
10a	Hardness-Rockwell		D785	—	—	—
10b	Hardness-Shore		D2240	D44-50	D50-65	D66-73
11	Thermal expansion	'b'	D696	100-200	100-200	60-110
12a	HDT 264 psi	°C	D648	—	—	—
		°F	D648	—	—	—
12b	HDT 66psi	°C	D648	100-110	108-115	170-195
		°F	D648	212-230	227-239	338-383
13	Thermal conductivity	'c'	C177	8	8	10-12
14	Tg-when amorphous	°C		—	—	—
		°F		—	—	—
14b	Tm-when crystalline	°C		105-115	120-125	130-135
		°F		221-239	248-257	266-275
15	Linear shrinkage	in/in	D955	0.015-0.050	0.015-0.050	0.015-0.040
16	Specific gravity	-	D792	0.91-0.93	0.91-0.94	0.093-0.095
17a	Water abs. in 24h	(d)%	D570	<0.01	<0.01	<0.01
17b	Saturation water abs	%	D570	—	—	—
18	Dielectric str.- short time and with 0.125 samples	v/mil	D149	400-1,000	400-1,000	400-500

'a' foot pounds per inch of notch: 0.125in specimens used.

'b' thermal expansion means the coefficient of thermal expansion with units of 10^{-6}in/in°C.

HDT stands for heat deflection temperature or, deflection temperature under flexural load.

'c' the units of thermal conductivity are 10^{-4}cal-cm/sec-cm^2 °C.

'd' water absorption tests performed with 0.125in specimens

TABLE 9B. MATERIAL PROPERTIES.

	Property	Units	ASTM	HDPE HMW	HDPE UHMW	PP-H
1	Tensile strength	psi	D638	2,500-4,300	5,000-5,600	4,500-6,000
2	Elongation at break	%	D638	150-800	400-500	100-600
3	Tensile yield strength	psi	D638	2,500-4,000	3,000-4,000	4,500-5,500
4	Compressive strength	psi	D695	2,800-3,700	3,600-4,500	5,500-8,000
5	Flexural strength	psi	D790	—	—	6,000-8,000
6	Tensile modulus	10^3psi	D638	135	150	160-215
7	Compressive modulus	10^3psi	D695	—	—	150-300
8	Flexural modulus	10^3psi	D790	120-170	130-140	170-240
9	Izod impact strength	'a'	D256A	3.0-4.5	No break	0.4-1.0
10a	Hardness-Rockwell		D785	—	R50	R80-102
10b	Hardness-Shore		D2240	D63-65	—	—
11	Thermal expansion	'b'	D696	70-110	125-130	80-100
12a	HDT 264psi	°C	D648	—	110-120	120-140
		°F	D648	—	230-248	248-284
12b	HDT 66psi	°C	D648	155-160	—	220-250
		°F	D648	311-320	—	428-482
13	Thermal conductivity	'c'	C177	10	10-12	2.5-3.0
14	Tg-when amorphous	°C		—	—	—
		°F		—	—	—
14b	Tm-when crystalline	°C		125-130	125-135	165-170
		°F		257-266	257-275	329-338
15	Linear shrinkage	in/in	D955	0.015-0.040	0.030-0.040	0.010-0.025
16	Specific gravity	-	D792	0.95-0.955	0.94-0.95	0.89-0.91
17a	Water abs. in 24h.	(d)%	D570	<0.01	<0.01	<0.01-0.03
17b	Saturation water abs	%	D570	—	—	—
18	Dielectric str.- short time and with 0.125 samples	v/mil	D149	450-500	650-700	550-600

'a' foot pounds per inch of notch: 0.125in specimens used.

'b' thermal expansion means the coefficient of thermal expansion with units of 10^{-6}in/in°C.

HDT stands for heat deflection temperature or, deflection temperature under flexural load.

'c' the units of thermal conductivity are 10^{-4}cal-cm/sec-cm^2 °C.

'd' water absorption tests performed with 0.125in specimens.

TABLE 9C. MATERIAL PROPERTIES.

	Property	Units	ASTM	PP-K	PP+ 40% Talc	PP +40% Glass
1	Tensile strength	psi	D638	4,000-5,500	4,000-5,000	8,000-15,000
2	Elongation at break	%	D638	200-600	3-10	2-4
3	Tensile yield strength	psi	D638	3,300-4,300	4,000-4,500	—
4	Compressive strength	psi	D695	3,500-7,500	7,000-7,500	8,800-9,800
5	Flexural strength	psi	D790	5,000-7,000	8,500-9,000	10,500-21,000
6	Tensile modulus	10^3psi	D638	100-150	450-550	1,100-1,500
7	Compressive modulus	10^3psi	D695	—	—	—
8	Flexural modulus	10^3psi	D790	120-200	450-600	950-1,000
9	Izod impact strength	'a'	D256A	1.0-20.0	0.4-0.6	1.4-2.0
10a	Hardness-Rockwell		D785	R50-96	R90-110	R102-111
10b	Hardness-Shore		D2240	—	—	—
11	Thermal expansion	'b'	D696	60-90	50-80	27-32
12a	HDT 264psi	°C	D648	110-140	180-270	300-315
		°F	D648	230-284	356-518	572-599
12b	HDT 66psi	°C	D648	180-220	260-290	330-335
		°F	D648	356-428	500-554	626-635
13	Thermal conductivity	'c'	C177	3.0-4.0	7.0-7.5	8.0-8.6
14	Tg-when amorphous	°C °F		- —	— —	— —
14b	Tm-when crystalline	°C °F		160-170 320-338	155-170 311-338	165-170 329-338
15	Linear shrinkage	in/in	D955	0.020-0.025	0.008-0.015	0.003-0.005
16	Specific gravity	-	D792	0.89-0.91	1.23-1.27	1.22-1.23
17a	Water abs. in 24h.	(d)%	D570	0.03	0.001-0.03	0.05-0.06
17b	Saturation water abs	%	D570	—	—	0.09-0.10
18	Dielectric str.- short time and with 0.125 samples	v/mil	D149	550-600	500	500

'a' foot pounds per inch of notch: 0.125in specimens used.

'b' thermal expansion means the coefficient of thermal expansion with units of 10^{-6}in/in°C.

HDT stands for heat deflection temperature or, deflection temperature under flexural load.

'c' the units of thermal conductivity are 10^{-4}cal-cm/sec-cm^2 °.C.

'd' water absorption tests performed with 0.125in specimens.

TABLE 9D. MATERIAL PROPERTIES.

	Property	Units	ASTM	EVA	PPVC	PPVC filled
1	Tensile strength	psi	D638	2,000-4,000	1,500-3,500	1,000-3,500
2	Elongation at break	%	D638	300-700	180-450	180-400
3	Tensile yield strength	psi	D638	1,000-6,000	–	–
4	Compressive strength	psi	D695	–	800-1,600	1,000-1,750
5	Flexural strength	psi	D790	–	–	–
6	Tensile modulus	10^3psi	D638	6-30	20-60	40-90
7	Compressive modulus	10^3psi	D695	–	–	–
8	Flexural modulus	10^3psi	D790	7-8	40-150	40-190
9	Izod impact strength	'a'	D256A	No break	0.4-20.0	0.4-20.0
10a	Hardness-Rockwell		D785	–	–	–
10b	Hardness-Shore		D2240	D17-45	A50-100	A50-100
11	Thermal expansion	'b'	D696	150-200	70-250	70-120
12a	HDT 264psi	°C	D648	–	–	–
		°F	D648	–	–	–
12b	HDT 66psi	°C	D648	–	–	–
		°F	D648	–	–	–
13	Thermal conductivity	'c'	C177	2-5	3-4	3-4
14	Tg-when amorphous	°C		–	75-105	75-105
		°F		–	167-221	167-221
14b	Tm-when crystalline	°C		100-110	–	–
		°F		212-230	–	–
15	Linear shrinkage	in/in	D955	0.007-0.035	0.010-0.040	0.002-0.030
16	Specific gravity	-	D792	0.92-0.94	1.16-1.35	1.25-1.65
17a	Water abs. in 24h	(d)%	D570	0.05-0.13	0.15-0.8	0.5-1.0
17b	Saturation water abs	%	D570	–	–	–
18	Dielectric str.- short time and with 0.125 samples	v/mil	D149	600-750	300-400	250-300

'a' foot pounds per inch of notch: 0.125in specimens used.

'b' thermal expansion means the coefficient of thermal expansion with units of 10^{-6}in/in°C.

HDT stands for heat deflection temperature or, deflection temperature under flexural load.

'c' the units of thermal conductivity are 10^{-4}cal-cm/sec-cm^2 °.C.

'd' water absorption tests performed with 0.125in specimens.

TABLE 9E. MATERIAL PROPERTIES.

	Property	Units	ASTM	UPVC	UPVC Glass filled	PVC/ Acrylic alloys
1	Tensile strength	psi	D638	5,800-7,000	9,000-9,500	6,000-7,000
2	Elongation at break	%	D638	40-80	2-3	35-100
3	Tensile yield strength	psi	D638	–	–	–
4	Compressive strength	psi	D695	8,000-9,000	8,000 13,000	6,500-8,500
5	Flexural strength	psi	D790	13,000-13,500	10,000-15,000	10,000-11,000
6	Tensile modulus	10^3psi	D638	850-880	300-550	300-380
7	Compressive modulus	10^3psi	D695	–	–	–
8	Flexural modulus	10^3psi	D790	300-500	700-750	300-380
9	Izod impact strength	'a'	D256A	0.4-2.0	0.5-1.00	1-12
10a	Hardness-Rockwell		D785	–	R118	R106-110
10b	Hardness-Shore		D2240	D65-85	–	–
11	Thermal expansion	'b'	D696	–	–	–
12a	HDT 264psi	°C	D648	140-170	150-155	165-185
		°F	D648	284-338	302-311	329-365
12b	HDT 66psi	°C	D648	160-165	130-180	170-190
		°F	D648	320-329	275-356	338-374
13	Thermal conductivity	'c'	C177	3.5	–	–
14	Tg-when amorphous	°C		75-105	75-105	75-105
		°F		167-221	167-221	167-221
14b	Tm-when crystalline	°C		–	–	–
		°F		–	–	–
15	Linear shrinkage	in/in	D955	0.001-0.005	0.001-0.002	0.002-0.003
16	Specific gravity	-	D792	1.30-1.55	1.50-1.54	1.25-1.35
17a	Water abs. in 24h	(d)%	D570	0.04-0.4	0.01-0.02	0.09-0.16
17b	Saturation water abs	%	D570	–	–	–
18	Dielectric str.- short time and with 0.125 samples	v/mil	D149	400-500	600-800	450-500

'a' foot pounds per inch of notch: 0.125in specimens used.

'b' thermal expansion means the coefficient of thermal expansion with units of 10^{-6}in/in°C.

HDT stands for heat deflection temperature or, deflection temperature under flexural load.

'c' the units of thermal conductivity are 10^{-4}cal-cm/sec-cm^2 °.C.

'd' water absorption tests performed with 0.125in specimens.

TEMPERATURE CONVERSION

In the temperature conversion table (see table 10) the center column contains the numbers that you wish to convert. If you have a temperature in °C that you would like in °F then, read from the center to the right. For example, if you would like to convert 200°C into °F, then find 200 in the center column and then look immediately right: you should see 392°F. Check this by dividing by 5, multiplying by 9 and adding on 32.

$$\frac{200 \times 9}{5} = 360.$$

Then 360 + 32 = 392. (or 200 x 1.8 then +32).

To go from °F to °C then enter the figure you have in the center column and look immediately left. 200°F becomes 93°C. Check this by taking away 32 from 200, dividing by 9 and then multiplying by 5.

200 − 32 = 168.

Then $\frac{168 \times 5}{9} = 93.$ (or 168 x 0.56)

TABLE 10. TEMPERATURE CONVERSION TABLE.

Centigrade	Starting Value	Fahrenheit
−18	0	32
−12	10	50
−7	20	68
−1	30	86
4	40	104
10	50	122
16	60	140
21	70	158
27	80	176
32	90	194
38	100	212
43	110	230
49	120	248
54	130	266
60	140	284
66	150	302
71	160	320
77	170	338
82	180	356
88	190	374
93	200	392
99	210	410
104	220	428

Centigrade	Starting Value	Fahrenheit
110	**230**	446
116	**240**	464
121	**250**	482
127	**260**	500
132	**270**	518
138	**280**	536
143	**290**	554
149	**300**	572
154	**310**	590
160	**320**	608
166	**330**	626
171	**340**	644
177	**350**	662
182	**360**	680
188	**370**	698
193	**380**	716
199	**390**	734
204	**400**	752
210	**410**	770
216	**420**	788
221	**430**	806
227	**440**	824
232	**450**	842
238	**460**	860
243	**470**	878
249	**480**	896
254	**490**	914
260	**500**	932
266	**510**	950
271	**520**	968
277	**530**	986
282	**540**	1004
288	**550**	1022
293	**560**	1040
299	**570**	1058
304	**580**	1076
310	**590**	1094
316	**600**	1112
321	**610**	1130

Centigrade	Starting Value	Fahrenheit
327	**620**	1148
332	**630**	1166
338	**640**	1184
343	**650**	1202
349	**660**	1220
354	**670**	1238
360	**680**	1256
366	**690**	1274
371	**700**	1292
377	**710**	1310
382	**720**	1328
388	**730**	1346
393	**740**	1364
399	**750**	1382
404	**760**	1400
410	**770**	1418
416	**780**	1436
421	**790**	1454